Comtesse Paul Pokorski, a writer of articles about astrophysics, has written this novel with a will to enlarge her acquired knowledge to a worldwide audience and share her way of analysing the link between Earth and space, making everyone aware of the progress of science in the field of astrophysics. Many concepts will be treated, such as the link of space to time that will be discussed, dark holes, light, elementary particles, so that any reader can understand that time has come to be conscient that we have to preserve Earth from any modifications impacting its atmosphere, pay attention to pole displacement modifying Earth mass repartition and thus changing seasons. The destiny of humanity is endangered on a much larger scale than we think.

To any woman: the world itself is founded on duality. We have to bring our contribution to science and be more confident in ourselves.

To Thimoty, victim of hate and nonsense.

Comtesse Paul Pokorski

THE LIVING WITHIN THE MULTIVERSE – BOOK 1

AUSTIN MACAULEY PUBLISHERS™

LONDON · CAMBRIDGE · NEW YORK · SHARJAH

Copyright © Comtesse Paul Pokorski 2023

The right of Comtesse Paul Pokorski to be identified as author of this work has been asserted by the author in accordance with sections 77 and 78 of the Copyright, Designs and Patents Act 1988.

All rights reserved. No part of this publication may be reproduced, stored in a retrieval system, or transmitted in any form or by any means, electronic, mechanical, photocopying, recording, or otherwise, without the prior permission of the publishers.

Any person who commits any unauthorised act in relation to this publication may be liable to criminal prosecution and civil claims for damages.

A CIP catalogue record for this title is available from the British Library.

ISBN 9781398495265 (Paperback)
ISBN 9781398495272 (ePub e-book)

www.austinmacauley.com

First Published 2023
Austin Macauley Publishers Ltd®
1 Canada Square
Canary Wharf
London
E14 5AA

Special thanks to Nicolas Irribarne who made me more self-confident about calculations.

To Paul

A Great Spirit and Mind

To Isabelle G.

Who has been placed into the light and has helped me recover the right way to the Truth, if we are allowed to believe it may be so. It requires so much virtue and Intelligence to even reach the first step of it.

To France and Henri, blessed hearts.

The future of humanity is the responsibility of all. Each of us has the possibility to find Salute and everybody has the choice of completing any holy actions.

We are all sisters and brothers and the actors of our human society.

Thus, we have to write our future, as a unique body.

BB

"There was once the Initial Illumination of the Light Spirit that made the Day and that will soon order to darkness"

In Nomine Patris Luminis

What was, will be again and what is to be again, will be forever.

The time has come to the advent of a durable Peace on Earth.

Eden will be the Original Garden again and the very one that had led men and women to sin, will be repelled to the night depths where he will rest.

The Bad Will Be Defeated

We are going to get closer to a story that goes back to a very long time and where we are going to, time does not exist. There is neither a before nor an after. There is what is. The universe and living beings.

At the beginning was the Verb, the unity into action: the Truth.

That is what appeared – what the first spark made visible. The truth, that is what was created. This truth is conceptual; it is in each of us. It is One. Unique.

It descended onto Earth with the fire sky, as fire has always been present at the moment of great truths. That is the way it is manifested to us, that the original energy bears witness to its presence.

Fire is not materialised; it is nothing else but Spirit. He cannot endorse any materiality but dictates World Law.

At the very beginning, the Spark came out of the obscurity where it was consigned to render all things visible, the original energy knew. It, only, is the truth.

Every participative element set into motion and since that instant, have never stopped going.

As Voltaire, the French philosopher, had once told, the astronomical clock had a creator. To carry the intention, a supra conscience was necessary.

The universe was created on the principle of the well-founded, where all the subparts of the Great All will warrant its continuity.

Stars and planets were split into specific tasks. Black holes would be the conscient keepers of the alter universe as they were charged with and of the Original Matter and of Dark energy.

Being gone right through by a blast and under the effect of pressure, they gave life to a flashing spark of vital energy, as a canon shot.

On Earth, life was. Everything contributing to its constitution, The Great All, will then concentrate itself on this minuscule part where fire and life would be established.

Here are we today, descendants of that original exploit. Incarnated conscient beings. Another extra conscience to ensure the continuity of the Great All.

But the One who was given incarnation to saw every element according to the law of its world. He began seeing what was visible without considering that everything had come out of the invisible or even think it could have.

Einstein established that our knowledge of the world was not reality but a tiny part of it. He rendered visibility to Invisibility. He showed that our thoughts originated limits to our visions of space.

How could we believe in an invisible infinite dimension?

Could we imagine time as not being our time?

Could we imagine that some universes composing the multiverse were not visible?

That there are some hidden worlds, as with the theory of the hidden developed by Vallee, other theories according to which there could be mirror worlds (Russian recent theory)?

Others that suppose the light split coming from an infinite source of light?

If I am sometimes doubtful about it, it is because doubt participates in the revelation of knowledge. It authorises access to some conceptual truths. Without the notion of doubt, knowledge would remain as it is and thus would not access Truth.

Indeed, have I thought that if I do not reconsider my certainties, I would not get closer to truth and the reality of the world.

Maths imposed itself as the right matter to define those realities and truths. As it would help demonstrate them and open new doors, even if we have in mind that an axiom can also be demonstrative of the Absurd.

We also have to keep in mind that there are cultural ways of thinking in link to calculations. We cannot interpret any calculation if we do not refer to the same scale, to the sharing of consensual signs and demarches. We have to adopt mutual references, but those references have no universal characteristics.

We have to conceive of everything with an enlarged way of thinking. As an example, the representation of a tetrahedra is a reality. But another reality is caught in the volume itself. As it occupies some space, we could also establish that it occupies some space into time. That it occupies an inner position that is an interval point into time. That it has a time reality.

At the very centre of the Singularities into the Dark Holes, there is an infinitesimal point where all the Star matter is concentrated. And there is also the horizon of events, that point, below which, everything is captured and trapped, including light.

Realities profile some spaces, limits, frontiers and even zones. Those realities are accessible via the formal truths of maths and geometry.

Another element to be considered is circularity. If between P1 and PO, a certain time interval has passed from the conscient point of view, in reality, an event has occurred. To go from one point to another, there can be a derivation of the point itself due to the incidence of electromagnetic waves and of the other alternatives enciphered in reality. If we consider that we evolve into time space.

Humanity and the universe, a transcendental intention or an immanent One? Insignificance of humanity? Singularity of human life?

Whatever one can think about it, we each possess the duality of the cosmic elements and integrate the experience of the Great All into our inner selves.

Listen to the celestial silence, to the ticking of the Cosmos, resembling that of your heart, and tell yourself:

I am altogether One and multiple. I belong to the world's truth.

It is more than ever necessary to adopt some principles of positive evolution, where all the elements of the Great All will serve one unique objective: the Universal well founded.

The time has come when humanity is confronted with its evolution.

Tribute to Indian Chiefs

When the last tree is cut down, the last stream poisoned, and the last fish eaten, you will realise that you cannot eat money.

Sitting Bull
We have to preserve our forests, for all these which cannot speak as birds, fishes, or trees.

Qwatsinas
We do not want riches. We want peace and Love.

Red Cloud
A very great vision is needed, and the man who has it must follow it as the eagle seeks the deepest blue of the sky.

Crazy Horse

General Summary

Introduction
Biography
Initial Elaborations
    What?
Glossary
Debate
General Presentation

First Part Axioms
A. Generalities

1. Is the System Intra-Modulable?
2. Is It in Name of the Adaptative/Evolutive System Function?
3. Monade and Supreme Unity
4. Could the Universe Be Differential and Conscient?
5. What Is to Be Deduced Then?
6. The Permanent Living

B. Studying Further

   1. The Notion of Good and Bad
   2. Interactions of the Living

C. An Impermanence

   1. Teleology and Destruction of Earth
   2. Can Eternity Be a Human Perspective?
   3. Time as a Retrograde Concept and a Recursive Wind?

D. Spacetime and Hypotheses

   1. Possible Distortion of Spacetime
   2. Which temporal dimension for our humanity?

E. Randomness and Determination

   1. Randomness or Determination?

Second Part Reasonings

A. Dark Holes

   1. Principle
   2. Which Models of Representations Could We Reasonably Induce?
   3. As Concerns Gravity Constants
   4. Development: The Multiverse and Dark Holes
   5. Attraction

B. Time Concept

1. Principle
2. Different Suppositions About Time
3. Models and Coefficient of Discordance
4. Is Time Possible to Define?

C. Space

1. Principle
2. Necessary Preamble to the Constitution of the Existing
3. By Referring Ourselves to Our Presuppositions
4. Proposition

D. The Universe

1. Principle
2. Developments
3. Other Developments
4. Referential Glossary
5. Other Suppositions
6. Other Suppositions

E. Motion

1. Principle
2. Other Suppositions

F. Calculation of an Original Source Point According to the Systemic Principle

1. Principles
2. Developments
3. Principle of Perennation of the System and Its Rebalance
4. Effects

G. Twin Aspect of the Universal System

1. Principle
2. Developments
3. This Subject Interests Ourselves

Introduction

The Universal domain is so vast that touching on the subject in a few pages would be an offence to its magnitude and its great power. There are so many elements to consider. To go so far as to say that I will have to change my mind on some of the ways I adopted till then to better appreciate the reality of those concepts.

As to the multiverse, which we have decided to call the Interverse, as it is interconnected with the other parts of the Great all, it could be questioned as the imbrication of several universes, which, acted according to the principle of circularity, could contain, in the core part of the whole Set, an autonomous, self-balanced and conscientised system, if we consider everything to be conscious.

The universe is One and multiple, sequenced, with also the possibility of sequencing itself. A multiverse.

Just as well, the principle of circularity could be applied more generally to matter, when matter coupled with energy are the factors organising the living.

Why could it be impossible to conceive that an elementary particle could not have any mobility intersystem? A particle can use any possible options until passing by an obstacle. It is altogether here and there.

It is, consequently, difficult to evaluate a defined position of the elementary particles unless adopting a measurement system, but any measurement does not reflect the reality of their functioning as the measurement considers the specific instant where the particles are observed, and this action modifies reality. Any right measurement is possible due to the mobility of the particles at measurement time.

From then on, each concept is an idea, an axiom something to demonstrate.

How can we feel sure that any supposed concept is logical? Is logic a concept to apply to reasoning?

We usually admit the idea of quantum randomness, as anything is as undetermined as the functioning of particles.

If I wrote that they obey to a cycle, that they are coming perpetually back or round to an original Point, it is not illogical but refutable if I do not demonstrate it. However, being not in a position to suggest a proper measurement system and being not able to propose a reliable instrument of measurement, even not able to proceed to a quantitative analysis, I can only suppose or theorise.

And as concerns the universe and the living, there are many theories.

Some plural universes, issued from the original light, which is unique. (Its manifestation into sub consequent energy integrating the living and animating the living, lead us to think that it is a linking operator.)

The coronal hole into the sun, impacting matter on a general plan. (we are made of matter and that matter is under the influence of waves)

-The Quantum Double: "Everything is originated from an action of DE doubling existing in the infinite smallness and in the infinite width." (The Doubling theory)

So many concepts and ideas that we are going to develop throughout our essay.

We can hardly represent ourselves as made of universal particles, acted on by some similar elements as those impacting the cosmos.

Indeed, if everything was designed to maintain the Balance of the Great All, our human level, we have to endorse some of the universal duality, as well as the impact of individual energy and define ours, according to a physical and mechanical point of view. We are not only made of flesh.

Consequently, we should think of symbolising any acting factor, as well as human energy, think of defining some body constants, the Cosmic constant R as affecting ourselves, the power of quarks, the acting pressure of spacetime, the exchange of bosons, the evolution of a lepton in neutrino...

Leading us to question ourselves on this postulate that sounds as an adage: Man feeling in phase with the world!

Without coming onto the ground of metaphysics, this notion is interesting, as it links the living to the universe.

They form a unique and same set with some notions of translativity from one to another. What is acted, acts in its turn, and what has been acted acts again on the original acting factor. Circularity or ternarity. Whatever the name. The principle is:

The feeling of following a line and going ahead towards an objective that permanently adapts itself to evolution, without thinking of a global Intention (I mean, from our human perspective), renders those sets only reactive to punctual actions, when we should observe and conceive projects in the long run by determining an Active strategy.

The interstitial exchanges did not integrate the notion of circularity. We all conceive everything around according to the principle of duality. For example, to Plato, Real Love was not a simple relation from one lover to the other, but to a triangle, the third element constituted by the world of ideas. The identification to the third summit of the triangle is a linking idea to Justice and Good. What, in true Love, defines attraction is that both converge to the loving of the Knowledge of Ideas.

Odd parity thus had to be conceived. Parity was not efficient enough and revealed that it was not operating when we started conceiving the first computer programmes. System operability depends on a third element. 0, 1, 1, as an example…A couple of figures cannot be operating.

Indeed, why should we go from here to there? When we could go from here to here?

There is Alpha and Omega. Omega symbolising the necessary balance and Alpha, being the cosmologic constant.

As you can see, we do not tend towards a goal but to the balance keeping of the Great All. That is a system design.

Within the universe, whose goal is the preservation of the System, men have to experience or experiment with universal energies. We also have to think of the human experience of The Great All in the human incarnation. He is an infinite part of the energetical Great All when conceiving human life as an entity. He can then destabilise All.

To get back to the notion of the two sets: the living One and the Universal One, we have to fusion the sets. One is the other. Think of Plato, and what we have just developed. Those two sets are connected to a third. A conscious knowledge of what is necessary to the system. In the sense of good. The link is active. And the system conservation could be symbolised Mo+ evolution, and as a function of the cosmologic laws, we could be authorised to think that we could be able to anticipate the moment when a split particle could come back to its original point Po, circularity as a cycle. The cycle is the result of ternarity. If we consider time to go and never get back on itself, we attribute it a move, a motion way. But we have to change our minds on that particular notion.

For example: 1)1)1) composing the multiverse with the Power 3.

The balance of those sets and the maths law depend on a principle that is not known. We have to elaborate a concept from a reality that cannot be determined on a specific instant, as we still think time is a determination.

Time has no before, no after. Time as a cycle. A new value is necessary.

I have interested myself in the word Dieu into French and saw it as an acronym: Domini in eterno Universu...In English, the acronym could be GOD as the Greatest of Powers that we have chosen to call: the Great All, a dominant energy. An initial energy. The light. The very one that was.

It is not to make the affirmation that it is so. But to show you that everything that was established as being defined according to a consensual reference can have a different value than the one that has always been attributed.

What is important: Changing our way of seeing things. At a time when energy is more than mass and celerity. At the time of our evolution, other spaces and other possible universes are opening to us.

Perhaps, will we be capable of defining a space time point, considering the derivations of the point? The acting factors on that point? To go to and always be capable of getting back to our departure point. Necessary if we want to travel into spacetime. This point is to calculate. Because we could go into another spacetime without being able to get back to the departure spacetime.

To integrate what we really are:

One and multiple, at the image of the universe. A representation of a tetrahedra occupying some space into time.

Think differently of a volume and change our point of view by redefining the height?

Because, when reasoning from a starting based point, we lose in accuracy.

Everything has to be thought of and reunited into a unique concept, integrating all the functioning representations.

To me, it sounds like a unique equation: that of light genesis. Master energy at its source. Reproduce a source of light so close to the original, but that is not fire. Fire being possible on Earth. We have to think of universal possibilities. Energy is vital to matter.

It will always be one of our major preoccupations. As nothing ever remains as it is.

A Greek philosopher had stated that what characterises time is not its value, or its determination but the fact that each change is motivated by a primal goal. An objective that has not found any mirroring in men's attitudes.

We have then driven our Earth without any global vision of a final point to reach or of an objective. Mastering and controlling can originate from a will to reconsider our attitudes. Will a council of wise men reveal someday itself as a possible objective to achieve?

Being wise is giving the possibility to those who possess wisdom, knowledge and intelligence to be the right guardians of an active future. Our future.

The whole set interacts and integrates in the name of the Great All. There are some subset systems constituted within the Generic Sets, but they constitute a whole set. Everything can be calculated from one of any constitutive elements.

It if we could be a part of All, in the Great All, a dream will be completed, this, of a durable peace when, as in the novel by Martin Luther King, the dove terraces the snake.

The law of the living has not to be ignored. The cycle rounds to ensure continuity. Ongoing process which, in order to last, has eliminated all that was not necessary.

Self-balancing, intersystem predation to maintain the system?

Whatever the word. The living is an entity. A global entity, persisting in different forms. It has to be thought of as a whole.

As to us, we would like to show or try to underline some predictive conceptual truths.

Here are some of the axioms that we would like to bring out.

It could seem ambitious proceeding so, but our will is to better understand, while keeping in mind that every incoherent supposition will be refutable from our way of reasoning, and that some axioms are not the evidence of the physical laws. They can be fake suppositions, even if generally admitted.

For example, the square of hypotenuse is not equal to the total of the squares of the two other sides. Plato demonstrated this in his time. For many reasons, but according, first, to the Retrograde Demarche and secondly, to the fact that it only applies to plan structures. But everything evolving into spacetime, curbs should be considered, volumes, then. Or solids. A friend of mine, Nicolas, with whom we were discussing the subject, gave the example of the Little Prince by Saint Exupery. If he had to replace a triangular metal part of his plane, calculating the square of the hypothenuse would be of no use to him. This would not correspond to the real measures.

It helps to show that nothing can never be evidence of the real functioning of the elements that compose the living. If everything could be demonstrated, then the Existence of the Lord…

To some philosophers, doubt was an argument in favour of the existence of A Creator. But, far from us the idea of discussing the point, we are expressing our ideas; they are not reduced by any ideology. They are placed under knowledge and intelligence, ways of superior wisdom. And guided by the interest of the collective, as well as the major stake linked to our universe.

Analysing the knowledge of knowledge and suggesting some possible ways. Here are some of them that we would like to be predictive or predictable.

- Any particle contains some Eo and infinite; they are the evidence that past acts on the future present. Present time is a conscient concept, not a real one.
- The anticipation of bridge point in spacetime, of which every factor would be predictable.
- The creation of an Alpha and Omega from the fragmentation of the Multiverse.
- The universe is fractal and there could be some specific decompression levels to go from one to another.
- Wormholes that could link two distinct regions of spacetime and that could be potential short ways to go through spacetime. Where could the bridge point be?

And some other subjects, some of them linked to morals, as in the chapter dealing with the notion of Good. But we do not initiate a philosophical debate; our knowledge in philosophy would not authorise such a discussion.

Our purpose is to suggest some conceptual possibilities. It is a novel at the image of Aladdin's cave, an 'Open Sesame' novel…

There are some more consensual conceptual subjects, such as the dark holes and the gemellar/twin Worlds.

The universe being fractal and where there could be some specific decompression levels to go from one to another? Wormholes that could link two distinct regions of Spacetime and that could be potential short ways to travel through spacetime?

However, these are theories that would be included in ours.

It shall be thought of as the visit to a house. There is an entrance and a door opening into the Grand apartment of the World. The living room where we live, the cellar where we stock, the room where we keep all that will be recycled...

During the visit, we have to be aware that what we see is what our conscience wants us to see and what culture has always shown us, but it is just some part of reality. We cannot see all the elements at the same time. And in the fraction of a second, what exists, could not be anymore and what would be then, could be there forever.

Earth is our common shelter that mixes functionality, usefulness and beauty. This is not our property; we belong to it. Of our quarks getting back into dust again, other elements will emerge. We are taking part in a system that creates and recreates permanently.

I have listened to some expressions, such as being atomised by the defender or having some crush atoms in common with...we have in mind this idea of belonging to a Universal order. Double-faced: obscurity and light. However, there are many intermediary nuances to go from one to the other.

If our living order belongs to a larger order, we have to consider it and choose the right way.

Each move has a direction that first does not appear to us as being such. It is in the energies of the world and does not have any influence on the Greater Order, as it is so small a determination. Except if humanity as a whole decided to serve the Great Order or the General Interest. Everything being conscientised to its own participation in the whole.

So are we.

And if this whole set, the number of interactions goes towards the preservation of the whole. It shall be the same with the dark holes that could restitute as much as they absorb to reach a neutral value, a Venturi effect pression/speed.

Everything is so interconnected that acting on one part of the set implies a consequent reaction of the acted on the acting.

The understanding of those elements could give confidence and joy when knowledge brought at large-scale proceeds of serving our intelligence in order to improve our intention. Understanding as a real issue.

Humanity and the universe obey to the notions of the Well-founded and of the General interest.

The whole system is organised with a Supra conscience: it balances, rebalances, and follows the goal of preserving and eliminating each threat.

The Great All is the absolute unity, the sum of all the interactions of all its composers.

If, nowadays, the hypothesis of a Big Bang is not admitted as corresponding to the reality of the birth of the universe, we can think that dark matter and energy played a preponderant role in it. Everything being linked to the impermanence factor typifying the Universe. How could we establish some cycles? Even Earth does not obey regular cycles...

Our desire of viewing the universe as stable (permanent) and of creating come maths formulas as well as some determinant concepts does not take reality into account. We have to include the recursive concept and attribute a discordance factor and random factors, whose incidence would be nearly minor.

We, generally, think that the universe started at a specific moment, but the constitution of it and the gathering of all the constitutive matter preceded the creation. As a mason would.

Light was the most determinant factor of all, as the dynamic principle of the Great All is supported by the elementary particles.

A particular mention to the photon, it is interesting to notice that it is accreted by the Dark holes when it has no mass. How does this deformation of the Spacetime occur and where does a photon accumulate within the Dark hole? Far beyond the horizon of events? Next to the Naked singularities?

Regarding the photons, Planck and Einstein established that there was a link of proportionality between the energy of a photon and its frequency.

The Einstein theory is the closest to the reality of universal creation by giving a plausible explication as to the continuum spacetime and the curb of space.

Universal space is not static, so it is never the same, and the temporality is relative. There would be no ancestor to our multiverse but on original point where time includes a recursive loop, integrating past and future, as the principle of the determination of time, of the Instant, is a reassuring concept but an inexistant one.

The universe creates and recreates.

Is it in perpetual expansion? Can we predict an evolution/involution according to an elliptic curve?

Is there a possible ending?

As it creates and recreates permanently and by examining the Multiverse as a system whose evolution would be submitted to some law, we can reasonably think that no Ending could be possible.

The end of the living is not possible, too, if we consider life as an entity. This entity evolves according to an arborescence, as in maths; it integrates new extensions as new possibilities that would be favourable system extensions. It is not really probable that the system should be regressive. It creates hyperlinks as extensions to extensions. We can think it is Hyper evolutive, mostly if we consider it a system with its entries and exits and the links of all the system parts to the others on the principle of recursivity and of the retrograde demarche.

Our will is to understand with you how we can imagine The Multiverse at our time. Every year, new theories or discoveries redistribute any cards we had and have in hands.

Like that, the Great All remains a little mysterious and does not reveal its secrets. It goes on surprising us.

Are there any limits in knowledge made as protection barriers?

Is humanity sufficiently conscientised to the multiverse?

As knowledge must open wise doors and convey a reflexion, lead to the intelligence of future choices to envisage. So that knowledge is not the property of some of us but could liberate from obscurantism. Knowledge must guide to the right way and not to a lost path.

The decision is not to be taken on the basis of any calculations that could be made but to be evaluated with much favourable consideration.

We shall keep in mind that:

Matter was gathered,

To build a shelter on Earth,

From dust, Earth was created,

to gather and unite the living and the universe.

Biography

I graduated from two universities. Lyon, where I was born and where I studied English and American languages and Aix in Provence, where I studied for more than eight years, specialising myself in the field of human sciences with a doctoral mention cycle in research in the domain of maths didactics and a degree in multimedia engineering at the University of Lambesc.

I started questioning the universe by analysing it according to the Edgar Morin theory of system. It greatly facilitates the comprehension of the universe, as observed from the angle of a system and it is easier to analyse. Edgar Morin is the director of the European CERN centre.

The universe and its conceptual principles, viewed from the logical point of view of a system, give some evidence that every element is acted in perpetual move.

It questions the place of humanity within that set and our future, the question of the multiverse intentions. Primal questions, if we think of the living as centralised and that universal events have followed a determination series.

I have no pretention and will not be affirmative. Studying renders humble. I have no certainties and science always leads us to review our ways of observation. It is more prudent to try understanding, as from a better knowledge will come out a better mastering of the elements and of the base concepts, in view of ameliorating continuously and adapting ourselves to any change. Developing our consciousness. Never forget being wise and open to Benevolence as regards the universe.

Or future will be a condemnation; we must look out for positive evolution in regard to the system. After all, we are evolving within spacetime. Within impermanence.

All that follows:

Any models, principles, explanations are our own intellectual property.

Initial Elaborations

What?

The living within the Multiverse

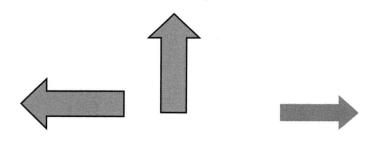

How?
Placing the focal on some key concepts
And deal with the knowledge of knowledge

> **What for?** It is the right time
> to feel conscious of the
> System Stakes in view of
> system perennation.

The living within the Multiverse

When? Where? A mini analysis to initiate a debate. This is a collective debate.

Glossary

Primal action: Primal Spark

Negative Alter: Differential Double

ALTER ALL: The other GREAT ALL created by the Original Energy and the Primal Spark

Fractal: Fragmented universe

Inter Time: Event period occurring between two sequences

Interverse:

Constitutive Set

Intraverse:

System intra system

Involution: Regressive move

Monades: Perfect Unity of the Great All

Isotopic Twin world: Hypothesis of a twin world

Multiverse: Name given to the several universes composing the whole set

Predaction: Evolution of predation

Source Point: Generic Entity

PTD: Point of tension/distortion

The Great All: The universe and the living, as a Unique Set

The recursive loop: the recursive move or the return on itself, the circular move made by a system on itself or towards itself. This is mostly linked to spacetime and time.

Debate

I was writing this novel, but the meaning of it was difficult to perceive. The reason why?

And at the moment, when I was thinking the book was fully written, everything became clearer. I had the idea. The reason why.

I thought that soon everything that was not visible would be made visible. The light will be inundating the universe, and the Obscurity will disappear. The sky will be a transparent space, as the light would have helped brighten everything.

Do we have to calculate a process? Will the secret of the Primal action of Truth be revealed?

The answer to my questioning was that our souls needed to be alighted. That our eyes only see a fake image of the physical reality of the world. Our reality is that light inside Plato's cave, bringing a feeble ray of light to puppet men, confined in a destiny they have been deprived of by the reigning obscurity that had provoked: the absence of knowledge.

With the revelation of light and by learning to consider recursivity, ternarity and circularity, they will be forced to view that the universal reality was different, and our human visions of the universe were truncated versions. The invisible (dark matter, dark energy, mirror worlds or equations), which was an inherent composer of the Visible, was not considered in any calculations. For example, a tetrahedra has many faces that we cannot see at the same time. The front representation and the back one...We admit there are sides. That this is a volume into the space. Do we believe sky to be One levelled?

Answer me: The sky is a plan surface? Is the visible part the only one?

With some correct calculations, the sky will appear as it is. It will be made visible.

Nothing will be opposed to knowledge. Knowledge as liberating us of our chains and not as suffering. It will fuse with the transcendental intensity of the Original Energy Eo.

It will help men feel part of the Great ALL; as we are energetic consequences and we have been conscientised, but we have to be conscient to the fact that the Great All is a Generic living Source. It is composed of any Animated element (in the sense of Soul) and the sum of all its composers.

Everything was borne by the Original Verb, which was the Primal Intention, The Original Action, giving light to the cosmos.

Could we then observe the cosmic interactions in another way?

And if we could imagine a bridge point into spacetime in a specific zone of the dark holes? With the possibility of calculating it, to anticipate all the derivations of the Point.

A bridge point opening onto another spacetime dimension in Ti(Time instant)

An invisible mirror world?

Physicians admit many theories: mirror Worlds, DE doubling quantum…you cannot imagine all that we know. Even a Hidden Valley where all the elementary particles would be concentrated…

We are not so far from great truths.

By widening our knowledge, everything will be visible and then, the Primal action: the Spark that unites the living and the Cosmos. Indissociable parts. To ensure the continuity of the All.

This is the 'time ahead, the following' of our humanity.

We have not considered the Great All as it is. We have omitted considering the invisible as coexistent of the visible. Duality. With a principle: The invisible and the visible are indivisible.

As to the other invisible part of the Great All, we have neither thought that it was influencing the whole, nor imagined its dimension in link to the known dimension.

We have just considered what was known, the Part of the whole that was consensual.

A part that is humanised. The real being that a human sees. But does he see all the REAL? Is it not the tiny part of a greater reality?

Imagine a world that would reflect itself to the infinite? What could we view of its reflection? A following of projected images aligned into a perfect symmetry?

What about the third image of the mirror? Is the reflected image of the first mirror and then consecutively of the second or the reflect of the first two?

It is the same as our universe. We do see only what we see with our telescopes, but what about hidden worlds, invisible valleys, some planets evolving in the shadow of another one, as the reflected image of the mirror or a Trick by David Copperfield, would it be visible? I well remember his trick with the train that the magician had made disappear to sight. He knew how to play with lights and how to use the directions of the rays of light and the reflections of objects into a mirror system (prism, dioptre). The magic trick was simple, in fact, if we knew about physics.

Change your mind about the universe.

We should first state that possessing some parts of knowledge is not in possession of the knowledge. We could adopt a technical approach to our observations. Design a protocol.

A system being designed by its limits and represented by the sum of its interactions, Systemic analysis could be a method or a way of analysis where we could be less inclined to get lost. We well know how to define a system and its mechanical functioning rather than interpreting an image.

How could we think of the link of the tree to the light without considering its shadow?

A painter would try to recreate the perspective, the shadows and when he paints, he interprets, as he reads the real with an emotion, a different perception. He sees other colours.

Let's interest ourselves in the painter's way of seeing the world.

How does he stabilise a colour with his/her emotions, how does he determine each volume, each line?

He/she paints with his/her soul. He sees with his/her heart. And restitute the information.

I am precisely thinking about the French painter called Cezanne and of the way he used to paint Mount Sainte Victoire. He designed a hundred paintings of it because the light was different each time and the reflects of light on the mountain constantly changing...

Try to think of him while observing the universe. See things with the painter's eyes, the film editor's sight. Calculating the sets of lights and the most appropriate observation positions to get a right return of the visible...This is a beginning.

And if this aesthetic side, if this primal link to light were constituents of the visual fundaments? See what is nice and not what is not; see only what we want in regard to our cognitive capacities?

How can I make you travel to the Mirror worlds; How can I tell you: See right there. Yes, but what?

Open YOUR EYES AND YOUR HEART to YOUR SOUL and think of what we have just said: the Invisible is also in our imagination. Imagination integrates the word IMAGE!

And see everything as an image projected to the infinite. Thus, an animated image will appear.

Is what we do not see what we call the unreal?

That is to say, we are not capable of defining the Unreal as a true part of our Reality. It means to us what is not real. What could we then think we really know?

Is knowledge so binary? White or black? What about the range of colour between the two?

Even between hate and love, there is a great range of emotions.

As for the universe, between A and B, there are many successive intermediary elements that keep in themselves a part of the infinite. As we.

The principle of circularity is linked to the principle of the successive action of the intermediary elements from one to another, a series of elements acting according to a hierarchy of positioning, and a speed of propagation. All these factors make a time interval different, even if the same parameters are implied. The action of the elements on the others modifies and changes all the time. But if there is some circularity in the system, could B get back to A?

However, it would not consider all the acting factors: pression, gravity, force. A will never be the same A as the original one. Ao. A will be the result of Ao. Everything is not permanent and evolves into spacetime that deformates itself at the passage of a Dark Hole.

How cannot we deduce that the appearance of stability is time illusion and that the definition of the living is that we are a result of what we subtract and a fragile composer to think as an Entity.

Life as a ternary cycle. Time between two intervals, not stabilised. Permanent interactions.

Why is the image truncated?

Because some could have intentionally modified it. Some could have deleted sequences for a film. We show what we want.

But why should I believe in what I see and say:

I can see so what I see, is.

And if the true question was: Can we only see what we want or anticipate seeing?

When looking at the sky or observing space, say: What could be there that I do not see?

A world that is not what I see, but the world that I THINK or guess.

The main key to access to the universe is the role played by the dark holes. It is a key opening onto an everlasting time, onto a deformable spacetime.

Imagine them as the backcloth. To what is it useful?

To play with words, we could think that there lies the real problem. But to the photographer, it helps to valorise the visible, enlightening it.

Then?

So, if the invisible renders the visible accessible, we have to think of what it is? Of what is it made of? What is hidden behind?

Black clothes dissimulate some sceneries that would not catch light enough or would not authorise the return of the colours scenarising the whole. We well think of each word used.

Catch light. Scenarise the universe.

Dark holes are so essential composers of the Great All.

Did they originate Eo? ORIGINAL Energy? How did it get through?

How was it canalised?

So many questions, but first, something important to do: Change our way of seeing the visible.

Opening, thus, to another way of reasoning, preserving ourselves from fake paths and errors. System error.

Admitting that another set exists, that could be the reflected, the quantum double.

Founding our knowledge on the visible truncated our appreciations of reality. Seeing it as a photo. Try to shorten, to rognate it, to resize it…

Whatever the angle from which we should see the image or appreciate its height?

Admitting the existence of an invisible, integrate it into our reality and think about the universe as a set of reflects, images, mirrors, hidden worlds...

Indeed, as men have not considered the Invisible as included in reality, they have elaborated a law of Fatum, according to which a series of random fabulous events had led to the constitution of life. A once-on-time story at a human dimension.

A fairy tale that well ended, as they had lots of children and life proliferated.

But can we honestly believe randomness is so effective? To get to such a result? And how can we explain that a set could mute into another one? It would be admitting that a system could bear its own generic conditions diverging from the original system and that the originated one would, in no case, have been anticipated by the system itself?

Randomness cannot bear any intention.

By definition, it acts without any objective and if we consider the number of necessary convergences...

The real issue is, which role did dark holes play?

The universe being predated and predating, the original set was submitted to some factors and vectors of perennation inherent to the system, but how could it create a different version of it?

Except if we think that it was guided by Superior Intelligence. An elaborative conscience to get back to math notions.

A Veil conscience ensures its preservation or continuity. Yes, it could be that: the system anticipated another development and if it did so, it was done with the objective of adding another favourable factor. Everything was concentrated towards an option: which is a probable determination.

The set originated from the original as an extension. Has it got an evolutive or regressive value?

We cannot know.

However, the system has designed some other extensions. We have already admitted this principle as a constitutive part of this reading: changing our way of seeing Reality, admitting by thus, the invisible composer to the Visible as an inherent part of it and attributing a value to it.

Thus, we could imagine the system to have originated many extensions.

Establish that what we see would appear under an inappropriate cognitive angle and attribute a variable to the angle positioning, a value to the invisible or a universal reference value 10^{99}

Never forget that what we do not see is coexistent in the real world.

When we analyse the existing, we speak of concurring energies, of forces, of a Set Dynamic, of an Intended first action.

The dynamic of the Great All is a creative one capable of evolving by creating what it is not and the necessary conditions to.

It is a system tending towards infinity.

At the same title that the set of real plus tends towards real minus, the system is differential. We are shown what others want.

Would there be a creative memory or, once the subset generated, according to circularity and the time concept, would it be constrained?

In that sense, would the system obey to a creative dynamic, not to die, by reproducing itself to the infinite? The system could think of a more conservative form.

And the more options there are, the more the system can last under the most reliable forms, the ones that better preserve the existing.

It would be a system with a speculative dynamic. Not a tentacular system but a superior intelligence calculating all the survival options. A hierarchy of functioning that we are unable to imagine. It redefines its parameters to adapt and evolve.

We will return to this notion in further development.

To end with this debate, it is important to consider that life would have built as an arborescence. The extension of an extension with many branches. Experimenting with conversative options.

Would we be one day the generic system of another and that everything has got one scheme: serving the Great All at best.

Knowledge is a means to wisdom, and the objective that we follow through these pages is to lighten a possible way, the one that makes us feel conscient of playing a role in the system.

Create by ourselves a new branch of the arborescence, as each of the extremities anticipates, on its turn, the preamble to the next existing, a future evolution or involution. At the image of the first Set, create a differential set that will reveal a favourable extension, or a renouncement to.

We have voluntarily included no calculations (a few), as they should have been verified in the first time and experimented in the second time. We have interested ourselves in a concept that seems determining for our future but whose objective is ambitious: Light genesis (the name I give to it), what we see as the source primal light point...

Everything is dependent on energy and issued from the primal action.

In addition, considering some concepts and formalising them was an objective to achieve.

These propositions, being linked to some axioms that could reveal incoherent, we have chosen to keep them by ourselves.

We have considered phi, the gold number and after much reflexion, the same remark appeared to us. Changing our minds.

As we thought, there had been only partial visions of that number and each had developed a different theory, adding a stone to the building (it is much in link to geometry and volumes).

No global vision leads to a truth. Truth being founded on evidence that is not made to be discussed. Some pieces of a jigsaw puzzle.

Each saw what he wanted to see and even the Absurd can be justified via the right axioms. But if we agreed to adopt a more global vision? By displacing our eyes from the centre point and being interesting into all the theories?

We have then combined every element and it converged to the idea that what could have been calculated that way was the bridge point into spacetime. Why?

Via the pentahedral?

It could enclose the necessary volume and the necessary reflections with its number of faces (5), its base, its duality. Duality of forms included in the form itself.

Pyramids may have a signification. They are integrative of many of the suppositions.

The Fibonacci Series: The derivations of the point?

Every calculation will lead to the same number...

What follows will help us develop our ideas and suggest a reflection around some themes, such as:

- Randomness
- Spacetime
- Dark holes
- Time determination
- Motion

Do not forget to think of what you do not see and that does not appear in our usual reality. Some unusual perspectives that our mind can open and our imagination envisage.

Do not forget that every element is conscientised: the conscience to maintain the system alive.

If humanity was an autonomous entity into cosmic living, it would then be a free electron that would not integrate the interdependence of All.

Then, each part of the set has a common subpart to the All.

Admit that natural predation of the living can create some different connexions, but they are anticipated.

The evidence is in the fact that the system maintains itself and is not purely adaptative.

A superior conscience within superior intelligence.

General Presentation

Do we understand what the light, issued from the primal action, is? Is it the Original Truth, initiating the creation of the universe, as an acting link according to a vertical hierarchy, where everything is determined, unchanging and permanent, or is it simply what the divine light wishes we could understand: acting, within a continuous link to the impermanence?

Permanence or impermanence, ternarity of a time system evolving as a recursive loop cycle; we intend to list some reflexion fields, suggest some possibilities, models, physics concepts and on the way, try to think of their potentialities.

We have divided this novel into different sections in link to some scientific concepts and have chosen to discuss, transversally of their ideologic and philosophical significance.

So, for the first time, our demarche consists of developing the concept in an empirical way, by analysing the pros and cons, and even the contradictions of each conceptual hypothesis.

And, in a second time, we get deeper into our first intention by proposing some calculations and theories of the scientific concepts that we have just developed.

We integrate the Systemic Model of E. Morin as the most appropriate to our models and theories.

Our methodology should authorise some other types of answers to our questionings, the implicated motors of our ideologies, and the analysis of the concepts and their human significance. The possible consequences and potential solutions.

We will be trying to evoke this notion of Intra space between the Spaces, with a personal definition of an objective time conceptualisation, evoking the sequenced fragmentation of time and space via the retrograde and recursive analysis, and will interest ourselves in the universe and what we envisage as the MULTIVERSE. We will name it INTERVERSE.

Is there a finiteness of time spaces and of the celestial living beings into a Multiversal predation system, where the living gives way to the living to develop itself?

So many questions to ask ourselves, to help us think of our human spacetime into the acting Multiverse and to think of the universe as conscientised, when it has authorised the birth of some living form, thanks to a series of successive events borne by a unique scheme.

As energetic entities, constituted of elementary particles, we will be confronted with the fundamentals of what we are in truth, within the universe, into the world and in ourselves.

And if faith has given me an interpretation of the possible functioning of the cosmic and physics elements composing the universe, it is more than ever the right time to preserve the Existing, our Humanity and all the animal and vegetal species that are sheltered onto Earth.

I had a dream, where the domestic animals that were wild once ago or just extracted out from nature were attacking men, because they had been driven too far from their natural habitat. This is because I had just seen a couple walking a Lynx outside my previous city; in the fields, it was walked as a dog. It impressed me much. Other animals and the Eco system, too, could be future enemies to human beings. We do want to master control. But who are we to do so?

We have to live in harmony, eat of the animals, just to satisfy some biological needs, not to kill them as trophies, as the killed poor results of pride. Wisdom must order men. It is an indissociable condition for the pursuit of any way that will be chosen.

The well-founded commands on any development.

We should also understand that within a SET, everything is interacted and acting in return. If one of the elements is modified, the whole is modified. If a set part is acted, S is acted.

We cannot go on to be selfish and irresponsible to our environment. Thomas Pesquet, through his videos from the ISS, launched an alert and showed that pollution of the ocean seas was a problem to solve rapidly. Too many plastics into the ocean seas, and the living chain is totally destabilised. To produce so massively, to destroy some ecosystems, to eliminate some predators as wolves, acts in disfavour of any harmonious development.

The objective of these pages is to give some empathy to knowledge. As well, widening our knowledge of the dark holes, will help develop the thesis of their huge utility to the Universal functioning and their great role in the stability of the Cosmos.

Are they invasive guardians invested in a general Interest Mission?

Everything seems to be acting according to an inner conscience contained in the living and according to the law of The Well-founded.

We are aware that anything can live and return to life again. That all is conceived according to a quantum recursive wind and that Life creates new connexions to Life, if examined as an Entity. With or without the human beings, The living Entity will remain.

But the light has always given a chance to humanity. Indeed, is it the case with the widening of knowledge. The fact of being able to travel through other temporalities could be a solution to the preservation: alter worlds or mirror worlds?

Quantum doubling? The bridge point into spacetime?

The living is a system in the system set. It eliminates any disfavourable connexion or extension. Via cycles. Anything that does not integrate the systemic notion.

To us, the time concept is not what we think it is.

We should bear our anteriority, and the future will never be projected to the infinite. As at the Cosmic scale, there is only one possible measure, that of continuity, as a consequence of T-1 and resultant of T+1. (We will be explaining that later.)

As nothing remains permanent, is it reasonable to think into the existence of the Instant, as it could be the resulting effect of an alternative projected into a future, materialised and existing in ourselves?

Two concepts compose the living: predation and reproduction. They could be demonstrative in themselves to their belonging to the Great All.

As human beings, we participate in the Continuity of the Universal System, whose depth cannot be imagined.

We are some conscientised incarnated particles. But we are acting in return.

It is of the scientific truths as of progress and new discoveries; they are always questioned. It is difficult to establish certainty. Everything can be supposed or supposable. Nothing is ever established as being so.

Medical progress seems to bring true answers against a specific disease or a virus that can be eradicated with a treatment, but then another appears. There is no whole answer to a virus.

We have to remain prudent with our current knowledge. One day, our solutions will be put into question.

Today, the role of dark holes seems in link to a universal stability or balance. They are predators and can reproduce in that sense that they absorb some existing matter, attract themselves and evaporate, enter in fusion, or repel themselves.

Earth, on its side, has evolved with absorbing a predated planet called Theia, after that one had impacted Earth. It gives us the hope that everything can be surely built with an INTENTION, and that it is not by a series of random events that it occurred?

There would be no logic in thinking that All will be the consequence of chance, having positively concorded towards the existing current living.

The system would then be evolutive and answer to some punctual actions, without any global viewing.

In mathematics, it is the sum of the interactions within the Set that characterises and defines the set. If we consider $E = Sa\ (FxxFxx\ldots)$

The system scheme is its continuity and preservation.

Those of we who will try to moralise behaviours, raising the conscience spectre of Good and Bad, will enhance a value that does not exist…

Does the predation exerted by each system to get it perennial reveal a bad acting?

No, as it relies on an essential principle, that is the one of continuity, and of the well founded as a result of.

The primal objective of a system is to ensure its enduring functioning and not to survive, what would suppose only punctual answers.

Except by thinking that a more important solicitation of the system would reveal a threat to it.

As the succession of events having constituted the system cannot be the result of luck, we can attribute some anticipation functions to it, according to an internal algorithm.

Indeed, we intend to try to understand the functioning of the whole Set. To formulate into some algebra simple propositions and to demonstrate that a mathematics Set is composed of all its elements and that any action influencing it is of a nature to modify it, as an auto-treatment solution is generated.

Moreover, at the instar of the universe, we should create a Council of wise men, able to define some objectives and orientate some decisions to make. It is more than ever a necessity.

We need an Earth design and not programmatic answers. Behave at the image of the universe, as some conscientised guardians, it is a NECESSITY to preserve our common good.

We have to think of our responsibility as being coexistent with the existing system. In the name of the general interest.

Then, I suggest a discussion about the rebalancing of the system from an original constitutive source point, which we will try to analyse through the potential interactions onto the living.

Does a twin isotopic space exist?

We also propose to grasp the seriousness of what could be the stakes of a better understanding of our universe from which we have to appreciate the functioning with a lot of meticulousness.

If the universe is fractal, it is multiple, offering a variety of possibilities.

The safety of the All will be in the global understanding and the Reflexion. We have to participate in the debate, being author-actors of our future.

As we have been conscientised to the world and the universe, have not we a role to play?

Cannot we affirm our responsibility within the All?

Writing these models and hypotheses was a moment of necessary thinking about our future. Suggesting, proposing?

There are many scientists who feel preoccupied with the preservation of the System, but publishing a book on the subject will authorise the debate to get out of the laboratories, put light on some problematic points and give an audience to it. A few do. WHY?

Our future is not the preoccupation of an elite. It is a common issue. All the parts of a set are essential to the set.

The Set principle is the consideration of all its interactions. The systemic analysis must integrate all the system exchanges: inside, outside and the very ones that are inherent to it.

How could have chance made so great things or done things so properly? Is it reasonable to think so, as therefore we are?

First Part
Axioms

A. Generalities

The objective of the following text is to understand as far as possible the link of the living to the invisible, the infinite and matter.

Of understanding that it is an undifferentiable set, linked to a universal set with unlimited frontiers.

Even if we can think this system to autoregulate permanently its transformations and changes, the acceleration of the internal interactions and their more important frequency could lead the system to autoregulate itself with less efficiency except by anticipating any event and predicting any external event.

1. Is the System Intra-Modulable?

It is not interchangeable at our time. It is not possible to create a Bis repetitive system of which the experience would be limited; we have to preserve the current one whose functionalities are well known.

The universe that we are going to call the current universe started itself from an existing one. It is false to believe that there was no anteriority to it. The dark matter discovered by Einstein, as far as I know, is at the origin of this transformation that can in any case be a creation; it is not because something is not visible that it is not.

Matter is evolutive. It is a true fact if we consider the system to be autoregulative. It changes according to its misfunctionings. It has an adaptive characteristic. The living One has evolved in link to its environment and any modification of its environment.

It has no selective intention. In that sense, it only selects what is necessary to its functioning; it is certain that it is confronted with some phenomenon requiring regulation.

But if it can autoregulate, it certainly has adaptative limits. The acceleration of the treatments is not the main issue when they only concern the system, but if those changes modify the external environment, it is the while system that has to be autoregulated and the succession of the adaptative regulations initiates a vicious circle of modifications.

In the image of the recursive loop.

If S1(S) is acted, it modifies S and S2/S3/S4...

We have to bear in mind that the change intervenes on all the elements composing the system, as it is a whole.

Indeed, supposing that those modifications are of an adaptative nature, can we reasonably believe in some limits to their continuity?

Some more questions are emerging:

- Is the treatment an instant regulation or is it made in the long run? An instant treatment or a repair? Global or programmatic?

[1]Is the system capable of anticipating any change or does it give only punctual answers. Considering the second option, it is obvious that an adaptive system would be auto limited. Bringing only reactive responses.

You understand that the more we will know about the Great All, the more our future will be manageable. We cannot remain inactive and at the image of the system, we have to bring out some solutions that will not be responses to events. Right answers.

We then have to analyse the system in detail and think about its globality. As the Great All is not the sum of all its parties but a unique PART.

The sum of impermanent parts will not equal the total. There are so many individual interactions acting positively or negatively in an additional or multiplicative way. The divisible does not seem to belong to the set, as it is expanding. it must remain commutative and observe a symmetry of functioning that we will be developing later, but if we subtract an element to the Great All, $S-1=S-1$, the set is not modified in its structure but in the sum of its possible interactions.

Why is this element subtracted?

The system GREAT ALL, having conscientised everything and being led by the WELL-FOUNDED principle, all that is subtracted, is not useful to the GA.

---

[1] Programmatic and scheme visions are parts of a theory developed by Jacques Ardoino, in the domain of education.

2. Is It in Name of the Adaptative/Evolutive System Function?

As regards its adaptation, the universal system is an evolutive one as it is predating and has the possibility of:

- Developing itself by activating or disactivating some subsystems
- Of adapting its needs in energy

It could seem evolving, on a random process without any predetermined intention, but the invisible dark matter of which the universe is greatly made of acts in a way to ensure its continuity.

In comparison to visible living, which needs to reproduce, to come to life and die, dark matter:

1. Dark matter ensures its own survival to maintain system continuity; it autoregulates.
2. Dark matter acts according to the principle of the living by accretion and rejecting or annihilating any elementary particles.

If a system is a predation one, it can act according to warlike answers and elaborate some strategies as an answer to any threat or attack. Even launch an attack intending to appropriate the characteristics of the absorbed system to evolve even more favourably.

This last alternative could describe an Initial Primitive System having not evolved, but a superior system **only** uses predation to:

- -autoregulate, as it conscientises predation in that sense that the system is interdependent on its great variety of actors and that they participate in the All Unity.
- -secure, as it destroys any threat onto it
- -ensure its needs in energy. Are they a necessity to whole living and to cosmic living?

We can well suppose the hypothesis of the Well Founded according to The General interest, to be more than a postulate, then a reality.

From Earth, the moon will be constituted (from the dust generated from the impact from Theia).

The universe will be permanent and the Universal balancing as the resultant of a constant redistribution.

It could be a perfect unity – a monade. In mathematics: the Supreme Unity and from a metaphysics point of view, an indivisible substance that constitutes the last element of everything. (Leibniz)

Thus, the role played by the dark holes is essential.

They are the basement or foundations the system relies on, and which help it being restored; they originate some waves, some light in some specific configurations of distortion of spacetime. They are guardians of a cosmic balancing, of a certain idea of permanence, ensuring the stability of a system to which ternarity and recursivity are basic principles.

Some waving sequencers of the Hadrons...the sequences they originate being an answer to specific needs in:

Redistribution

Regulation

The supplying of energy

They are matter sequencers, setting up some vital algorithms. Is it to be thought of as a repetition of the sequencing or successive sequencings?

3. Monade and Supreme Unity

If the system was purely additional, each subpart of the system would be unique and independent; it can then be multiplied. Suppose to expand to the infinite: $1 \times 1 \times 1 \times 1 \times 1 \times 1... = 1$

It is spread into a multiplicity of subparts, according to the principle that the Great All can be divided only by itself, and thus, the multiplicity of all the Great All parts, divided by their own figure, equals 1.

The great All is indivisible, and the difficulty is in the fact that it should be imagined as composed of subparts, as those subparts are indissociable from the ALL

4. Could the Universe Be Differential and Conscient?

The Great All or the ONE, is indivisible only by itself what makes One. It is One because it cannot be divided otherwise than by itself.

It gives evidence that whatever the number of the GA subparts, they constitute a same Set, only divisible by itself, Supreme Unity.

We could think that the figure 6 sequenced 666, would, on the contrary, be the representation of a non-identifying system, including no UNITY and being opposed to light. However, divided twice, it equals 1.

We will get back to this notion by evoking the evil twin *particles. Divided by 2, we get 333, which is in favour of circularity and Duality Plus + or Minus – of the elements. The particles: electrons, neutrons, protons...have all AN EVIL TWIN (this comes from a recent discovery). Possessing a similar mass to normal particles, they present an opposed electric charge.*

New evidence that to reach the perfect unity and aim at ONE, we could pass through a dark stage as components of the All, divided into two entities, incarnated and spiritual, to access unity.

And that the universe itself before being the current one would have passed through a dark stage, a necessary preamble to the existing. A cycle could be induced and the system, even if continuous, alternates both sequencings.

5. What Is to Be Deduced Then?

Would the universe be divided by itself in a binary way. Each planet has a double that could be its alter in the Great Alter All?

The concept of a twin isotopic system. How can we think of such a system?

If we can think that the minus sign, according to its math definitions, could be retrocursive to T-1.

If earth had a quantum double and referring ourselves to the concept of the recursive wind, it can, then, be authorised to think that life is a possible too. Indeed, by relating ourselves to the second part, it could be a matter of the living in T+1, the present being a resultant of two time lapses of which we cannot presuppose the permanent state, when everything is in move. It supposes T is not a real physics value and that time is conscientised. It cannot remain immobile and that, in any case. In other terms, living in the negative alter is what we are at T-1.

In addition, if Earth shelters the living (that we know of), we cannot hypothesise nor an exceptional phenomenon, nor a succession of random occurring. Randomness cannot be quantified, whereas all the elements of the set carry an algebra value and have a physics reality.

6. The Permanent Living

How could the living be defined?

Is it by representing to ourselves any element or body in movement and carrier of energy? As regards to this definition, the universe would be a full constituent of the living, having its proper autonomy and each planet would be a living entity in the same way as human life.

But we have to bear in mind that the Great All is a conscient system, capable of autoregulation, according to the immutable principle of the living: nourishing itself.

- In a primitive way, of subsisting
- In a more evolutive goal of development in relation to the spiritual.

The living is so inscribed into a wind that it is not reasonable to apprehend it as a chance. We have to examine the living from the angle of the retrocursive loop, continually evolving so that to refer to the reproducing Original A-1 towards A1.

B. Studying Further

1. The Notion of Good and Bad

Far from us, the idea that the world will not be submitted to any rules. If the universe seems connected to some notions of common well-founded, the human being has always sought limits but in vain!

Guided by personal interest confronted to the collective one, it has always been a challenge to men and women to balance both, in view of harmonising personal and societal life, and we have to tone down to this notion.

There is, from the beginning of humanity, a subordination link of the governed to their elected representatives, elected or not, to their directors and so on...

From then on, the private interest or the sectary one, prevails on the Cultural or community sense of belonging, as Humans have not founded a Human nation or Earth Nation, but a multiplicity of countries, cultures and religions.

Some interests in linking to some regions of the Globe oppose more or less dominating cultural groups.

The notion of Bad is being thought of in terms of non-pursuit of the Intelligence or in being beyond the reach of Wisdom.

In the name of world conflicts or wars, leading to the death of millions, to organise the protection of a specific geographical zone, or supremacy of a commercial nature, human beings have followed an objective of Self affirmation, that flatters his/her ego and ensures better omnipotence. He has privileged its membership to humanity. He should have globalised and thought of the universe and of any element with some kind attention.

Our Earth's main problem is to have not established any governance structure of a Pyramidal form, considering, at first, the Common interest of All men and a human notion of the Well-founded: the survival of all. But human nature is predation and human beings well integrate the notion of domination into the image of the pack of wolves.

The second problem is that of the ego.

This is a fundamental difference to establish. The universe predates to evolve. Human beings to ensure the survival of the dominant group. Darwin used to evoke the notion of the survival of the fittest, not in the name of All, or in an objective of securitising the others who are less feeble. Human beings ensure their control over others.

Moreover, nothing has never been aimed at. Every response has been elaborated upon as instant solutions to some events. Earth has been guided by an intention of its own and not by the intention of the GA. It has tried to free its chains from the GA and affirm itself as an autonomous, independent part of it.

From then on, if the general interest commands on the reset of a system, is it a bad acting, if it made to preserve the whole existing system?

It will be answered to... Who? When? How?

2. Interactions of the Living

If the living is parcellated into a multiplicity of forms, we have to notice that it constitutes a complex System inside which some successive retrograde interactions are made.

The living is an ongoing process. It turns on itself at the image of the planets and of the living reproducing in the infinite, in a wind. From the living, a new form of living is born and so on...The living is never to die.

Even after being absorbed by Earth, Theia went on interacting through the fusion of the two nuclei.

Conscient and conscientised alongside its evolution, this system has become more and more complex.

From this complexity, a paradox is born.

Can a complex system treat more numerous interactions?

Which evolution to Earth within the living could be envisaged?

All that is composing the living is due to evolve, to transform itself.

Will understanding be a key to some solutions?

The universe is in perpetual expansion. The moon is moving ostensibly away from our Earth and the will Sun, situated nowadays, at some right distance from Earth, remains at a favourable place to extend Earth living.

When chemistry, gases, temperatures and energies are to be questioned, everything is of an unstable nature.

The only lesson to be learnt from Earth evolution is that in spite of many cataclysms, the living has transformed itself. It can take different forms, all issued from an original one.

How can men and women perennate the living form they represent?

And what is the interest of a community more and more numerous and of which, the evolution could prejudice the continuity of the GA?

Indeed, is Earth condemned by the very species around which Earth living has constituted itself?

This would be one of the first vital paradoxes of the system. A paradox that could have already been anticipated by the system.

But, if it was not the case?

Humanity would have to make some decisions about its extension.

Thus, we could be confronted with the worst of all alternatives.

To auto destroy a part of the living to preserve it or doing nothing and taking some more time to make the right decisions.

Running time in the interval could play a disfavourable role against the system. It is information to be treated.

System alert?

We have no idea of the decisions that could be made.

Remain confident in the primal intention, when we think that we have been conscientised to the GA, to save ourselves: God helps those who help themselves!

We could also reasonably think that the GA, having attributed us a Conscience and FREE WILL, could be the only decision makers regarding the preservation of our species. But will the GA remain confident in humanity as it becomes more and more autonomous, and as the scientific progress could endanger GA Balancing?

There are too many divergent interests. It is, as if in mathematics, we had a multiplicity of sets that relied on some subparts of the other sets, functioning in an autonomous way. The common subparts only link human status.

You have understood that there is much to think about the subject and to play on words, we are confronted with a very DARK matter reflection...

We will have to answer to: Who? What for? When? How?

Each of us will try to save himself/herself and his/hers. The issue is that of everyone.

Our objective, through these pages, is to bring some consideration, initiate a debate, and possibly understand by opening up to the largest number of people as possible.

The tiniest parts of the All are GA parts. The human entity is a system in itself. It must seek the most harmonious and consensual response. Doing at best!

C. An Impermanence

Heraclitus of Ephesus had developed a theory according to which permanence was appearance. Nothing ever lasts but flees.

It is a theory that well integrates the incessant changings and the stakes of universal predation and that well explains the fact that the living is constantly renewed and able of reprogramming itself.

We have already evoked this particular subject of auto-adaptation and the subject of a programmatic system.

We have, thus, to turn to the question of the different scenarios of the destruction of planet Earth.

1. Teleology and Destruction of Earth

If we observe it from the point of view of the Common Well Founded, there could be many scenarios representing a possible threat and as well, the One when Earth could endanger the universe, in other terms, when humanity will represent a threat to the Balancing of any element.

Until then, the evolutive stakes have been favourable to the continuation of Earth living, but what do we know of the System Scheme? Of its evolutionary intentions?

Human beings are not constitutive of the living; on their own, they are one of the constitutive elements.

On this account, the elimination of human beings will be withdrawing one stratus within the living scale. But the living envisages from the angle of the system, would develop itself in a different way. The favourable extension of the arborescence will be put at first.

Predation is essential to the balancing upholding and to the whole system balancing. It is altogether a socle and a control lever. In that sense, it contributes to development and pursuit.

Nothing is ever as permanent as it seems, in fact. Every element is modulable and alterable, potentially into a process of destruction.

Would Earth be protected in the name of the well-founded that we have already evoked? Why should it thus be privileged? By a source entity or an indefinable invisible?

We must admit that impermanence is a concept in itself, difficult to apprehend as the world seems unchanging. We forget that we are interacted with by the energies and the elements composing our universe.

Dark holes and the invisible matter are as necessary to the cosmic balancing as air and to our lungs.

In addition to being a conscience, they are a breathing; some insufflate life to the universe as they evolve according to the expanding of the Multiverse.

Multiple destruction scenarios of Planet Earth could be envisaged, but we do not wish to establish a list of them all. Our objective is to demonstrate that the world is based on a primal intention that perdures. What Jacques Ardoino calls a "Visée". Transferred to the universe, it would be a Scheme borne by the system. This hypothesis contradicts the idea of a re-procreative wind, when our goal is to show that our universe could be reinitialised on some different bases. Would the ending be a destruction or a RESTART action?

« I will do new skies and a new earth.»

2. Can Eternity Be a Human Perspective?

Are men and women the only ones to have a conscience of time?

Is the universe relied on a sequential organisation, an algorithm of expanding?

An evolution/involution at the image of an elliptic curve, with an elliptic foci, that would be a point of possible decrease and of a programmed reducing of the universe (that is, ours).

If we recognise that everything seems in link to the notion of Well-founded or Common Good, the intention could be the survival and preservation of the existing?

In that case, Earth would be spared, humanity eradicated?

We envisage the universe without any possible ending, as if the multiverse had known a beginning, it would not necessarily come to an ending. We think everything to be cyclic and permanently renewable into a perpetual loop, a recursive one, the permanence of which being potentially established in that sense that a cycle is a repetition of some known actions. How could we imagine that the cycle could be broken?

If the spark which illuminated our world would extinct, in other terms, if the energy necessary to our continuity was suddenly blown out by *the same breath of the Spirit that initiated Human life...*

We are observers of the death and coming to life of many stars, planets, of some impacted planets as Theia and we could think there will be no ending to Earth? That our universe has no ending? If it were borne by a primal intention, a first scheme, it has NO ENDING; that is reasonable to envisage it according to that angle.

We think time, the universe, in link to some reasonings brought to a conscience whose inner scale of values is a subjective scale, by our human experiences, that cannot include universal predation as a useful function to the system continuation.

In itself, bad does not exist; it has no ontological existence. It obeys the law of expanding and of conscientised development.

3. Time as a Retrograde Concept and a Recursive Wind?

Starting from the concept of impermanence, our way of conceptually envisaging time includes but no notion of present time, which, to us, is erroneous?

Time is what goes from T-1 to T+1 on a straight line, of which the numerous sequences well integrate the value of a modulable interacted space of which we cannot imagine that it can be static at any sequence. By integrating a notion of thermodynamics, we think that the move made into a time laps produces an energy of which the simple action onto the environment modifies it. Nothing being ever static; how could we then think of determining an Instant?

At the image of the cosmos, where every element runs towards…

Time is not measurable; it is just sequenced, divided into cycles.

Can we admit the existence of time acting in parallel to time?

A theory of the recursive move in tension, of a time wind including two sequences. A couple of odd parity (-1, nought, +1)

We have no idea of a time that would not go, if we were put into any a vessel and enclose into it. Or prisoners of that vessel, would we reveal infallible to time…

Our lives include some cycles and periods.

And if we imagine the living as an entity, it has no ending.

D. Spacetime and Hypotheses

1. Possible Distortion of Spacetime

Is it reasonable, even logical, to presuppose a distortion of spacetime? The access to a dimensional space integrating another time value? Access to a dimensional space that does not obey the same geometric designs?

Our idea of spacetime is that a determined move, continuous, cyclic not of a deformable and malleable space, within which gravity would be modified.

It could be thought of as an Intraverse, a space between spaces, copying the principle of two relied sets, having a surface in common, authorising access to the other set in a bilateral way. A bridge point into two universes?

Well, we are at the stage of presuppositions, of propositions. We do not want to convince ourselves of the true nature of these spaces. Or sets.

2. Which Temporal Dimension for Our Humanity?

Into which dimension are we situated?

Can we establish a scale of value so that we can make some comparisons and adopt a symbol authorising those calculations?

Some ways of reasonings open us onto some interdimensional perspectives, to a specific Bridge Point into Spacetime, notions that we will be developing in the chapter dedicated to dark holes.

Dark holes seem to be the most accurate road to this fractal and fragmented universe.

Through some hypotheses of mirror equations, of the vision of a mirror world, of that Point where a special distortion of spacetime could be made. (to authorise travel into space dimensions)

We have to think of time and space as a unique set and only consider spacetime.

Has the multiverse been created on a transcendental intention?

Dark energy and matter, through the contribution they bring, their permanent interference into the cosmic balancing, are a possible right response to a lot of the questions we ask ourselves.

E. Randomness and Determination

On the word RANDOMNESS: This is a word we created. As we have admitted to the existence of a system selecting the best options, this term sounded more adapted. Chance is another concept that has nothing to do with the notion of an organised system seeking the best solution or resolution.

1. Randomness or Determination?

There is a fact we must be sure of.

Our planet cannot maintain itself into a stable state within a multiverse, or our universe, which is constantly in move.

Do you know that so long and long ago, it was Jupiter that was standing at a right distance to the Sun, and one day, Earth took it or exchanged it (?), to occupy the right place at the right distance from the sun!!!

There could be many suppositions:

- The universe follows an established plan
- The universe is a consequence of randomness
- Probability as a factor of determinism

What can we think of the stakes induced by these three possibilities?

We cannot reasonably think that randomness would do things so well and preserve the existent living, constituted after so many successive accidental events and do not want to anticipate or reinitialise by ourselves. If this hypothesis were confirmed, we would have to create a solution by ourselves to preserve any form of earth living that we want to.

In view of this objective, we would stand in the obligation of ameliorating our knowledge of the universe to open onto a field of possible options. With this hypothesis, humanity would be facing decision making on a survival issue.

What are the choices?

The preservation of the elite who runs the world? The scientific and managing elite? At the expense of any minority?

Placing human beings face the issue of their own continuation would be equivalent to considering some power stakes.

Except if we have kept the idea that we have developed a Council of wise men and women.

OR, considering everything was acted according to an established plan, whose objective was the centralising place of the living within the universe, and by attributing it to the Primal event of creation, then, as regards this logical consideration, remain confident as the living should be preserved. Any element that is concorded to create everything around it. A type of logic of the well-founded.

The universe would be perceived as conscient and obeying the principle of general interest. It would go acting with the creating primal intention: the development of the living and its continuation. Does it mean that humanity would be preserved?

If our species is the negative resultant of this arborescence, of which we have established that it created any farther extension that could enlarge the adaptation options of the system, and could we think of an extension that would not sound enough positive and would be erased by the system?

As to the third option, there would be some stages of probability, from the most to the least reliable.

A minima, we will have a chance that our evolution will go on according to a programme or plan without any bias but according to a plan that has already calculated the possible factors of variability and discordance.

In other words, considering that a projectile launched at a more or less high speed is going to follow a trajectory that is more or less predeterminable, we could then elaborate a line. Even knowing that an unlisted imponderable could change the line of the projectile.

According to this supposition, the idea that there could have been a natural predisposition for the elements to evolve is to envisage within the three hypotheses.

This reasoning applied to our human position within the universe, the singularity of life, and the unicity of humanity places us in a situation to define some future choices.

Either Human living and the universe were elaborated on a Primal Intention and in that case, there is a meaningful signification to our humanity. It would have been designed by a conscient intention into a conscientised universe. The intention would be immanent. Humanity would have been created as a differential alter, in the hypothesis according to which all is binary and includes a twin part. As for the elementary particles, which have an evil twin. (We have developed this point). Everything is built on the principle of the chain or spiral into which the tiniest parts are necessary to the energy production of the GA and its feasibility.

Humanity would be a co-creature, a useful double, in an intention of well-founded where men and women would have to reinitialise the All by themselves.

Or Humanity is of an exceptional no-intended meaning, included in a more global Entity that is *life* and Humanity, having not been borne by any Primal Intention, it must accept to define its future by itself.

In that case, the elite would elaborate on a plan for our humanity. That plan, would, without any doubt, be borne by an intention that would not conscientise the whole set but the only part of the human or Earth(?) set. By not taking the globality of the creation into account, even because of a misunderstanding comprehension, the solution would not be a long-lasting one and would have no chance of saving the earth living. In totality.

It would aim at sharing the wealth and power of that elite, guided by its own interest and not by the general interest.

Moreover, every project that is not animated of a conscience or an intention cannot bring peace. It would give priority, categorises. It is not of a universal nature, but of a sectary nature.

The objective of these pages is to bring a reflection, a contribution to a debate.

We must thus be open-eyed.

Everything is linked to our way of observing the invisible and considering our humanity.

The multiverse and the living. One stake: Ensuring and grasping Eternity to humanity.

Remember this sentence from EINSTEIN:

Chance is God walking incognito.

Second Part
Reasonings

A. Dark Holes

1. Principle

A lot of presuppositions have been made and some facts seem established as the distortion of Dark holes at the passing of matter and photons and the fact that they absorb some matter.

Is A part of that matter is restitute according to the principle of auto balancing, so as to the sum of the interactions absorption/restitution would have a neutral value?

Would any imbalance of the inner system of a dark hole be necessary to the exercise of a point of pression/distortion, creating an unequal repartition of the mass, thus facilitating the passing of a fulgurant beam of light?

What is the link between this passing through the beam of light and the original light?

Is there a reference of internal value (a bottom value or a limit) contributing to the mass balancing; Would there be a Mass constant?

If this option was confirmed, it would help us envisage many more hypotheses, but we would have to observe the evolution of a dark hole according to a referential and from a set of initially recognised consensual values.

How can we start from an unknown reference to define any known values? The dark hole under observation would be compared to some others to establish some statistics laws of a qualitative and quantitative nature to deduce any possible functioning model, as well as some purely putative rules. This would be an initial point to any calculations.

It seems, then, appropriated to write that a dark hole is a part of the Multiversal system? Indeed, it could even authorise access to it?

How are the elementary particles of matter that are absorbed and distributed within the dark hole?

And how can we explain that photons that have no mass could be absorbed too?

Dark holes are one of the most complex subjects and among the most interesting ones to consider. For that very reason, we have chosen to treat it at first and give priority to it.

If we were capable of conceiving an initial theory, we would dig a hole into spacetime (to play on words), the key access to the doors of the multiverse seems, indeed, necessarily originated by dark matter.

We can suppose that for an unbalanced of some Mirror worlds and then logically of a twin world, remaining invisible, would find some explanation via the analysis of the system that we chose calling INTRAVERSE, which is a constituent of the Dark holes.

To access these mirror worlds, there are many potential sketches. We think it implies some phenomena of light reflexion, and of an interface of two open mediums opening onto a closed one. In this title, we should explore the Descartes Laws so that to study how, according to a specific angle, the refracted rays in the same plan as the incident ray and according to the fractal principle would not be refracted into medium 3.

The access to dimensional corridors, provoked by a conglomerate of dark holes and, according to some convergence conditions, depending on a random rhythm, would open to other dimensions. At least to some other one.

We present that there are some corridors where gravity, its absence or less impacting, would authorise *gliding*, that is the appropriate term, and melt into an interdimensional passing by space.

Everything lies in a better understanding of dark holes.

We have supposed a distortion point that we called PTD, a point of tension distortion that would play a role in a very aleatory way, as a consequence of the convergence of many factors.

We have established this point as a logical consequence of a conglomerate of dark holes and to their:

- attraction/repulsivity
- attraction/destruction by disparities in the volume masses. To us, it lets a print. This is a mark of which we do not know if it can have the qualification of matrix. Can it originate from any calculation by starting from an elementary constituent destinated to absorb and originate some residual energy?
- attraction/fusion

To optimise the possibility of this point of distortion, the necessary conditions for a unbalanced repartition shall be instantly initiated. To re-establish the original balancing, the mass, whose centre has been impacted, emits a force so as to fusion, repel or destroy the other conglomerate of dark matter.

The resulting energy is authorised by the contradictory forces impacting the trajectory; a corridor forms itself, a hollow one, where energy can fulgurate. This canal authorises the circulation of some photon energy. We have specified into the introductory development that it requires some waving sequences of the Hadrons.

It has to combine some converging or diverging factors and vectors. Whatever. In the presence of another conglomerate of dark matter, matter adapts a reaction. It is the sine qua none condition to the liberation of a fulgurant beam of light of the same type as the one that originated at the origins of the GA.

Can we deduce that this instantaneous reactivity, this arm system, functions in a random way?

We have precedingly induced the principle of universal well-founding.

This would be the principle at the origins of the GA constitution.

As these matter, conglomerates appear to be predatory and to act in conscience; according to an expanding or creative principle, they therefore have an acting intervention. They are some systems of the primary set that have created some arborescence to each of its extremities so that to anticipate its evolution and, in the image of the whole, create a differential set that is a favourable extension.

2. Which Models of Representations Could We Reasonably Induce?

The mass of dark holes is of an inconstant nature, as they absorb and repel matter.

They are systems occupying some evolutive space, with a variable density, whose impermanent factors create an instantaneous imbalance, which causes pressure in a point we have already called the point of tension/distortion. It is difficult to localise within the dark hole. To symbolise it, we have chosen to envisage it as a fusion point between the two masses. This supposition could result in a third conglomerate that could justify the thesis of mirror equations and an alter world of an isotopic nature.

This point of tension would push in a specific area of the mass, creating an internal deformation of the mass, a sub consequent deformation of space at the same instant.

We could assume that this Point of TENSION DISTORTION would be situated in two points as functioning points of an electric circuit with points in A and B in the circuit, and a differential system initiated by each dipole as a function of the tension OF each of its electrical circuit terminals.

The medium and the density of the medium seem to be key factors.

At the moment when that push occurs, will there be a force at stake and that could also attract some non-massive particles as photons?

Could that point function as the point of an electrical circuit, with an intensity circulating as a function of the tension at each dipole: exit/entry. Two points, A and B, the photons would keep their energy potentialities and would exercise it at that point of tension, depending on the variability of the internal system. Photon energy is inversely proportional to wavelength. We must take this principle into account as another variable.

Moreover, two particles x and y of a different nature are agglomerated. How are they dispatched according to a structural algorithm?

XY agglomerating. What maintains x from y during a time laps?

Do XY fusion into another agglomerate and the sum of those agglomerates constitute a conglomerate, a sum of interactions that are devolved to create a misbalance of repartition into which the point would act as a partition algorithm?

Would that point be an original source point?

The entry door to a universal Multiversal dimension?

- Point of Tension Distorsion ?

—

1. Geometric representation: Their representation into a trumpet type like volume. If we concentrate on their volume, it sounds as if they were as follows:

Donut
Aspect when
Observed
From an opposite side.

They function as a closed space where time would pass within a subpart called Intraverse, in a universal sequence by making a recursive wind and where time instant Ti is only *determined* at the instant of the fulgurant beam of light. We can suppose that it composes at a common point, a timeless space, a dimensional door and that the push made at a parallel time would authorise the access to the Intraverse, a Multiversal entry door.

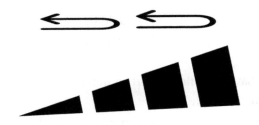

1. Given their appearance, we can suppose an unequal repartition of the mass with a deformed structure centre, the solid particles integrated within, being driven by some dispersion forces, according to the theory of the London forces, where at each inhomogeneous distribution, an induced dipole moment creates itself that can interact with the dipolar induced moments of the neighbouring particles: a force is exerted between them.
2. This force that is created at the PTD would have resulted in a distortion of spacetime. The length was modified by the tension force exerted at its free extremity. Given this element, we would add a hypothesis formulated at the end of the chapter of a

ternary regrouping with a neutral revolutionary (in terms of rotation) mass authorising the light transfer of both.

3. As Concerns Gravity Constants

Could we reconsider the calculation of the constants of gravity established long before our knowledge of relativity? Of the dark holes?

It seems that this constant could be useful to calculate but not a constant appropriated to a universe that is in permanent move.

And that we could attribute a more accurate value closer to the basic principle.

And create a calculation formula that does not *determine* the instant or a density. It must consider spacetime fluctuations. We agree that there is no question of determining a reality, as nothing ever fits into reality.

Every element is of a permanent inconstant nature.

G does not exist in itself, not in a term. We agree on everything to be circular and ternary.

When the pilot is in total weightlessness, no gravity is exerted on him.

It is exerted on a unilateral sense, and it does not make of this principle something simpler than the permittivity of free space, where it authorises the circulation of energy.

It should be recalculated into a more infinite value, which will consider the recursive quantum process.

4. Development: The Multiverse and Dark Holes

If we admit what we have seen precedingly that there is a multiplicity of universes, it presupposes:

A transitive relationship from one universe to another.

Defined as a relation of sequential type within the last and first of the universal elements S=S1+S2+S3...

It is at the very instant of their absorption/fusion that the energy is generated with fulgurant instantaneity. And it authorises us, calling it:

The Theory of the Cannon Ball

The light that will fulgurate is of universal nature, but we chose to give it the denomination of Original Energy, as its instantaneous development would be in link with the GA. It is so instantaneous that between its creation and the moment it is expulsed, there is no intertime, no time interval. It is quasi-instantaneous.

- Supposition :

To originate this light:

The integral formula is called light genesis, and the formula hereafter is partial.

The global formula includes some mesons, joules, a mechanical acceleration, some isospin baryons, an emission of leptons, a W13 pulsation, a gravitational potential, the alignment of the inertial centres, a Stefan constant...

Of course, there are some other principles at stake when light fulgurates.

However, values are difficult to determine as they are submitted to aleatory factors, including the principal characteristics of determination/undetermination as we do not know in which measure the randomness of these occurrences of concurring factors at a specific instant is not a provoked one.

The energy would be generated by the friction of two masses. This is what explains our model of this instantaneous beam of light by using F, opposing the move of two elements in contact.

RE=(mi1+mi2) Xm Xc/E0

X is the mass number. C= lightspeed. N moles representing mass m. Energy resulting: RE

E0:

F 1 + 2

But it is just part of my formula.

The formula englobes all the parameters. The values are difficult to evaluate.

We should admit, de facto, in preamble to the production of this energy:

- The necessary alignment of the two inertia centres without which they could not fusion.
- They would be too unstable. In spite of their huge volume, they present a density state of an unstable nature.
- Always one gravity centre. But when the two dark holes attract, we can presuppose the transfer of the inertia centre during the fusion time, probably

external to the two masses during a short instant, provoking a modification into the gravitational waves external to the fusion medium.

At the instant of fusion:

- The x rays, produced by the absorbed matter and the electromagnetic radiation, are ejected into a deployment of energy more intense due to the encounter of the two masses and the dedoubling/coupling of the two systems, even into a third or of a third. It seems logical that the systems possess the internal capacity to integrate any external system and to absorb it. However, dark holes have a capacity of absorption/resurgence of matter and transform it into energy.

We shall admit the idea that:

If we have thought that the creation of the Multiverse was the result of random events, it could depend on an anomaly of the gravitational field (according to the definition of it) and of a consecutive modification of the trajectory of dark holes, of their influence of action onto the systems which are then, reorganised or restructured into one.

We shall envisage the dark hole system of functioning. Think it is a necessary preamble to the existing. That they have been the 'carriers' of a primal intention.

This anomaly seems pre-determined and pre-calculated. It is necessarily linked to the nature of the masses in contact. Those dark holes possess the necessary elements to the constitution of the cosmic living.

We cannot speak of an anomaly but of an undetermination, in the sense that this factor envisaged as aleatory, is a constitutive factor of the system; it contains in itself the capacity to absorb and resurge into another. However, everything is constitutive of the SET.

But, the GA cannot be the sum of all the set included parts within a predation system that dies, regenerates and creates in permanence. It is *continuously processed.*

It does not generate a similar system. It creates an extra system that will help to the general functioning of the set of the GA. A dark hole can thus participate in the living constitution, but it is a reversed system where everything is translative and commutative. The system law requires a ternary and wind functioning. An auto-wind, making a constant return towards the original source point.

What is important is to think of the universal system as a link with that Original source Point, a generic light Entity, at the origin of the constitution of the Source Points composing the universes. Will they be localised into the dark holes?

AS they are a constitutive existing necessary to the GA and anterior to the original source point, they would have constituted.

We then tried to calculate the formula of the expulsed energy that would be close to Original Energy.

That beam of light of original energy cannot have the same mathematics declination as it is original and cannot be similar to any known measure. But any calculation on the Energy will take Eo into account and each calculation has to be in link to the first measure of the Energy that we shall be valorised (maths definition).

Energy and matter decline into different variations. The energy produced at creation has been an impulsive inner composer of matter.

We imagine the link between the Vital connection to the source point and the connection of human beings to spacetime. Humanity is linked to the universe, this one opening a gate onto it, placing this open space in a preponderant position to the Source Point.

The universe would have pursued a central objective and constituted itself around the living.

Sketch:

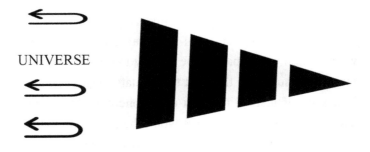

POINT SOURCE

The source point is a generic entity.

It generates a set of light rays into a beam. It is a starting occurrence to the source of light into the multiverse.

So, two dark holes would fusion:

Another hypothesis more difficult to modelise and calculate would be the hypothesis of three Dark holes emitting some instantaneous light. Which probability for three sets of dark matter to fusion?

They could, in the presence of opposite forces, evaporate or repel.

We formulate the hypothesis that they fusion, but that one of the three sets of dark matter is not visible, as regards the theory developed before, that is to say ternarity of the concordant systems:

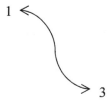

These three sets integrate a mirroring effect.

Adding the hypothesis of an Intraverse and of three sets with a common subpart to the totality into one of the masses that would authorise the light to circulate towards 3, 2, being in a state of instability, with an incoherent (Phys) medium density. One of the three sets would be a set authorising the transfer, a point of energy rewind. A rewind mass.

Its function would be to accelerate the transfer of energy into 3.

Three possibilities could then appear:

5. Attraction

Fusion/Friction with the creation of a point of tension/distortion and a force in M3, probably at the crossing forces F1,2,3 generated by the resistance to the move induced by the two masses in contact. With the probability of a mirroring situation into M3, in case three masses are in contact, regrouping three mediums (Ternarity). In other terms, it is the only one of the three hypotheses that would open on to the creation of an Intraverse, as instantaneous access to the multiverse.

Fusion/Division: The forces at stake provoke a repel through which the two masses resist mutually at the very point of tension/distortion. The masses distinct themselves to follow some trajectory proportionally exponential to their impact force under the push of the two masses at the PTD in view of opening on to a Hollow space where energy will be emitted into a beam.

Distortion/annihilation: The two masses annihilate each other under the intervention of a forces system of the push exerted at the PTD. It could depend on the position of the masses and their alignment at the perpendicular one to the other? A force impacting the other into its centre? In this case, the optimum alignment of the forces would be an angle action around 45 or 90).

These are three suppositions.

It should be preferable to create some animated sketches, to place any forces at stake, examine how the pushes could be exerted and where the torsion point would be created. Declining this sketch into some parameters that I cannot include, as I do not master the necessary tools of animation.

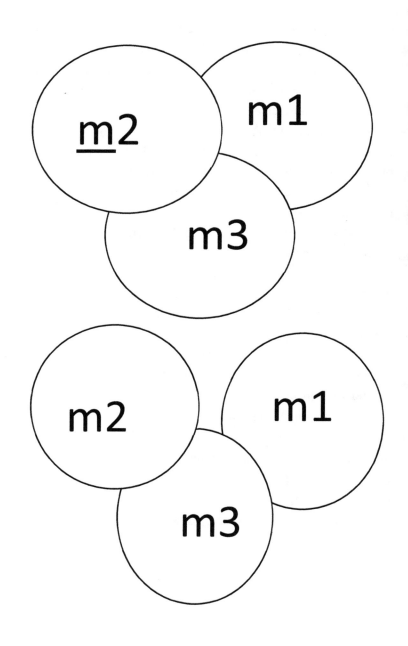

Some representations of the three sets symbolise their common subparts.

The first sketch shows:

Some subparts to m1 m2 m3: m1/m2, m2/m3, m1/m 3, and a common subpart to the three, but it is in m3 that the PTD would be situated.

The second sketch shows:

No common subpart to m1/m2/m3 but subparts between m1/m3, m2/m3 as m3 is the main transfer factor.

Which of these sketches best symbolises the forces at stake so that to create and develop a medium where spacetime is modified and acted?

To access this mirror world, we understand that there are some reflective phenomena or interfaces of two open spaces onto a third that is closed.

Given this, it would be interesting to study how, according to a definite angle, the reflected rays in the same plan as the incident ray and according to the fractal principal would not be refracted in m3.

This leads us naturally to the hypothesis of an M3 medium, which is a hidden resultant; it is not visible HOWEVER it exists. A third medium independent of the two others originated by the circularity in the mediums. We still admit that there is no gravity constant or refraction constant. The nature of each of these physics actions includes a retrograde demarche, an inversion, a circularity that makes of the instant, a coercive temporality (F). It constrains us to think of time into sequences as it emits some winds. As the square of the Hypothenuse synthetises the calculation of two sections (as an example, we have previously discussed the subject of the formula), our two mediums would bring evidence of the third. This is because of the principle of circularity.

B. Time Concept

1. Principle

Nothing is more conceptual than time. It is a scale, a measure unit. It is not obligatorily a consensual factor, as human time is of a mechanic nature that can be transferred to algebra.

We do not know the value of one day; the value of one day cannot be defined exactly with the following ones.

We admit that, as long as this value is not defined as such, it cannot be a referent reliable value.

When it is a question of abstractions, we go out of the real, which opposes the existing and has no value of possible comparison.

They are representations.

We think that temporal data is out of our human reality; it has to be rethought according to a value and calculation scale.

There is no strategy for our purpose. We cannot give evidence nor tell it is not. Simply give another aspect of reflexion on some concepts that have enough probing adequacy, so that we can test their accuracy to truth.

Human time has a calculated supposed value based on the repetition of natural elements and on their cyclic aspect.

Are there any time intervals, some Intertime lapses according to the circular principle, indeed, more specifically ternary? And is past living time the time of the instant when it is supposed to intervene after the time of thinking and is a mentalised representation before the action time?

As it is a concept, we are going further into the notion of concept to understand the values so that to define truly time.

- We understand that there is a discrepancy between thought and action. Do we project it before materialising it?
- In other terms, does the awaited action precede by its conceptualisation the resultant one?
- If yes, it supposes a notion of time that we call Intertime where a mental repetition of the future action will be materialised and our decisions the resultant of some actions projected into time. During that time laps, does time function in a recursive wind?
- Impermanence as an acting factor – some individual factors of motion. We are acted on and acting on. These are sketches of that intertime (inter the time) that we are going to define into the following development:

<u>One is defining the conceptualisation according to which Intertime would be acted, the other is the result of the inductive materialisation of the first. The mechanic time is uncapable of materialising correctly motion time.</u>

When we are thinking of an action, does our brain project an action that may not be realised? But is this projected time an acted one? In other terms, face to an alternative, our brain envisages two possibilities and are these actions at a hypothetical stage mentally projected? Do they include any time reality?

- Only one part of the alternative is going to be materialised, but during thinking time, both exist.
- Have these choices any influence on the individual and how can he or she master the permanence of his/her state face to the constant impermanence of the situations to encounter?

These are all the points that we will develop hereafter.

We could establish a vector of permanence VP. Any individual being is supposed to evolve within a constant of physics values and into a world that we presupposed to be stable, but which is in a continuous move.

They would be submitted to Motion factors (FM) issued from the historical parameter, inducing a sub-consequence developed into stages, but into a multisession or a multimodal order.

If: individual/time = VP + FM(H)

That is, to establish nothing as ever being permanent. And the necessity to create a new symbol evoking the functioning of time into a spiral or loop. VP(FS) factor of recursivity.

If motion does not oppose itself to permanence, it is because the system is circular that it obeys the recursive law, from Point One to Point Three.

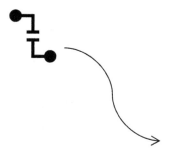

2 reflected to 3.

I would have liked to find some other symbols. To better express the principle of a transfer in a point towards another one.

Or:

It does a Loop to its anchorage point as an auto sufficient system. That point is the source point.

Instant T1+ consecutive of T-1, but at T-1, the line has been modified. Inst1(cons 1) and we should include any new symbol. That circular system acts in parallel:

This loop or wind (electric circuit) is dependent on the factors In/out of the instant. The instant is the resultant object of an impermanence constant factor, and it is never the same according to when we refer to any other motioned (waved)Instant. I at instant T1 or I at T2 (a sequence) is not the same because of intertime.

If we do not consider that notion of intertime, we will never be able to travel spacetime. Intertime is not measurable. It is submitted to constant variations, to some external factors of which celerity of the light at Instant T or to the wavelength interdependent of the streaming effects, of earth revolution and intra-Earth vibrations.

I repeat that what is discussed is motion into space-time while keeping individual characteristics and understanding that any action has an impact on factors that are only waves and energy. The instant is divisible and divided, submitted to some influence with an aleatory character.

- Then, the intertime of a second can never have the same time value. It does not liberate as an elastic and is not liberated proportionally to its primary tension. Its impermanence can be calculated. It would be submitted to some factors that, despite their impermanent state, would always be the same.
- We could adopt the predictive value of a second. It can be the *carrier* in itself of more space than spacetime itself. Because of circularity.
- The first day, not constituted of constant values and not referenced to any others, can in the fraction of a second having been constitutive of the world. The time connection functions into systems inter and intra tangibles. Then, could we suggest what follows: F(ex) + F'(inter)+F(intra)/Light+ RT (earth revolution) + VS/T

These external factors are determined by earth vibrations detected by some seismic tools, sea streams, gravity (itself in link to T). A unique unknown parameter can appear; the universe divided into a multiverse and of which we cannot measure the spacetime value, even if we attribute a human value to light and sound, how could they be permanent when perpetually interacted by the external elements.

Can we truly believe in the permanence of elements submitted to waves and inconstant energy and whose values change permanently. When I wanted to calculate a bridge Point, all the derivations of the Point were so important that we could think, for example, of going into a temporality, but how could we get back to our original one?

That was the issue. I think that some of us have all the parameters. The values…There lies the question.

We could adopt an algebra model as a reference and accept it to be + or − beyond the referent value.

However, humanity likes constant principles, which is not at all a universal concept.

Would the universe be in perpetual expansion?

Evolution implies involution and from the creation phase, we have an alternative: either an evolutive universe or a regressive one.(It concerns our future.)

The instant: A wind circular circuit.

- Individuation

Fx/F-x

Individuals are the resultant object of a permanence vector and of I1. Considering that we are floating into spacetime:

Inst 2/inst 1 + 1 inter time

I1 + inst1 = I1inst1
I1inst2/I1inst1
Circular system acting in parallel

1=3   2

Instant +1, consecutive of -1, but at -1, the line to 1 has been modified.
Instant1 (cons -1)
Then, we suppose:

1. No determined character of time.
2. An instant is an abstract concept, a calculation value. It has no cosmic reality.
3. Is the individual interacted? It has no existence into a determined reality, into a permanent temporality; it is in perpetual motion at the title of the energy he is motioned by.

We think of each theme in a unilateral way and from one point to the other and vice versa, but the recursivity vector of this time loo on itself leads us to believe in a circularity that we can only fragment into sequences at the image of time itself.

I1/inst 1=I1inst1=Inst3

The instant after the instant is a consequence of -1 and when we get back to -1, the line has been modified by its projection onto 1. Then, inst1cons (-1) is submitted to the circular cycle of time, but that time is reflected to time as a mirror, what we could express as the versatility of time, that it can go from one to another. By projecting ourselves in an 'out of time dimension', as the intertime is always in the time, we would always be in our time circuit.

We shall admit what we cannot judge of the value of a first day dedicated to the creation and of the value of the other day that would not be better calculable.

The whole universe is into our humanity. We are made of atoms and quarks (I quote those last ones to please my father as his service of research discovered the quark: IN2P3), of less infinite particles and we evolve into an environment made of atoms and of which the Ions bearing electrical charge divided into protons and electrons, whose composing number is different, are symbolised into cations(+) and anions(-).

But why is such a binary functioning?

However, could we not imagine the interval between the negative and the positive or a type of arborescence with many possible sequences that can go from -- to ++ or +- or -+ and decline them so that to get some more accurate calculations in comparison to the real we think like.

2. Different Suppositions About Time

We cannot conceptualise, apart from a binary system, as the one we know. Day and night, white and black. A reductive system that does not consider the differences of the internal components.

E. Morin conceptualised the system. He has designed it according to some internal and external influences that act on it, as well as he has imagined some permanent in and out entries and exits that we can presuppose, being not able to anticipate them all.

However, between those two values, there is another one. Unknown. This is because it is necessary for one to be constitutive of the other so that they can form a sequenced set. How is that sequence activated?

Which value to time?

If we consider it to go on in continuous, we shall admit a continuous timeline, some spacetime where time observes motion linearity.

- Are they universes formed as parts not included in the composition of GA, of a GA? a subpart constitutive of an All and then the GA?
- Will time be the same? Is there a regressive time? A progressive one?
- Has it got a final extremity? a beginning and an ending? Or is it that energy described by Plato as divine and that always gets back to its original point by doing some loops onto itself.
- That was what rendered the retrograde move incalculable, even by calculating with the square of the hypothenuse, this one being unable to verify the exact correspondence with the form triango-circular evoking recursivity. The formula by Pythagoras applies only to plan surfaces. We have already evoked that point.
- It was already a theory about recursive quantum.

3. Models and Coefficient of Discordance

How can we modelise such a principle or reduce it into algebra, as the quantification of time holds an aleatory factor as it is interdependent of internal or external phenomenons which are not predictable, always by considering the systemic principle. How can we adopt a model to symbolise the return motion and has it even got the same duration, how to symbolise this recursive action integrating in itself the extra time fraction?

Probably less if this return implies a time curve. We do not know how to conceptualise time. We have only conscientised and subjectivised it.

Each prevision must be opposed to reality. The possibility of testing some results authorises an estimation of potential measures, some errors being also possible.

Even if the difference in what is awaited and obtained is weak, if what we get is not what we are awaiting, the response is discordant. Some type of dichotomy between the real and the supposed real. If we can observe the results by considering the sum of the squares, the average square of errors presupposes a non-conform awaited by referring to Plato.

As to arithmetic or time conceptualisation, we have but no certainty except by calculating with a coefficient of discordance, integrating the aleatory factor and a difference with the awaited, could bring some sufficient evidence of the calculation.

That coefficient could be called the coefficient of discordance: Cd and the difference = or -1 could constitute some evidence about the awaited. It remains a weak difference, then a convincing one. We do have to determine some aleatory variables to make the results more accurate.

4. Is Time Possible to Define?

It seems that humanity has the a priori of a life purely natural, into a cycle linked to that of nature: with a beginning and an ending. But the vital cycle makes of us, some parts of a living whole, it makes of us some out of the time beings, as the vital cycle reproduces itself to the infinite through the atoms (quarks) we are made of during a period of time. Even if that period is definite, it makes some participative beings into the world energy. Our atoms return to the world and symbolically 'All are from the dust, and to dust all return'.

It is then supposing that time is cyclic and returns to its original point.

If it returns to the original point, it gives evidence of the reprogrammative character of time. Auto-loops into a finite process with two extremities (In/out)

It is not a versatile system in the sense of adaptative because as it adopts a cycle: it is not transformable but purely processible, that is to say, that can be acted onto.

It is a concept submitted to some variations, but of which the constant is that IT IS perpetually the same.

Viewed from this angle, this concept observes some constants of permanence and some variations submitted to motion factors that can be as conceptual as the presupposed constants.

Ideological time is society's time. It is a time with a cultural identity objective that has got no relative objectivity. In that sense, it is a fragmentary point on a straight line.

IT is made to section humanity's stages into historical periods and events. It has quasi-sensible value. It sectorises and adds a non-scientific variable.

- *Within the time concept, has the instant an alternative value?*

Does time exist according to two-sequenced loops integrating a hollow space, nought? Can we think about it as a motion into a 'time tenaille'?

For example, at the image of the cosmic world, it would obey to attraction/retraction, setting some regulation forces with a double way despatching.

In other words, we could find another spacetime. Time within the time.

By thinking according to this principle, we would obtain some 'time depression' at the condition that:

adopting a move in tenaille, two circular inversed moves that, by gravity at stake, would create a point into a depressed space. A sinusoidal function of time?

To finish with, is it a depression zone with aleatory functioning? Or can we calculate its pseudo sequence? I began having that idea while considering the rocket Apollo, which went to the other side of the moon. To me, it brought the question of time.

By gathering a series of such factors as:

$$Xy\ (m)^2$$

Y= mkl0 -

K unit of thermodynamic Kelvin-

X= E0-1

And does this move in a double recursive way adopt a vibrating move, into a stable point at the image of the tension point by Coulomb with a constant Lo of tension?

Tel $V= (T-1/T+1)\ /m$?

From the beginning, we have specified that we would make some hypotheses. In view of a better understanding of the elements and concepts.

C. Space

1. Principle

The notion of space has disinhibited our imagination by giving free reins to our representations of the universe, so difficult to apprehend from the point of view of its Hugeness brought to the human scale, that it has had even some divinatory aspects, where the planets should influence human destinies, in link to their positioning at birth time, trying thus to align human destinies onto the gravitational (I read vibrations were used) forces or the cyclic changes of earth position to the other planets. Can the universe influence our actions?

It would be necessary thus to conscientise the unreal by giving up objective the universe, what reduces it to nothingness, from a factual view. It is to human beings, to subject the universe and to attribute it a role/object.

However, space is what authorises animated living, and it cannot be reduced to nothingness.

Is the world issued from an original spark of light?

If we admit the existence of quasars, we shall admit:

1. The notion of a universe in permanent accretion of matter and, consequently.
2. The constitution of some light versions of the dark holes, of energetic agglomerates (electromagnetic emission).
3. A unidirectional Interverse functioning according to an alignment to a source point.

Indeed, we have imagined a prototype machine that would authorise to understand the recursive action of the source point and its effect; it would be a mass with some properties of synertial symmetry.

Could we then postulate the hypothesis according to which the source point would be at the origin of the Multiverse? Then, according to a source point.

Indeed, could the multiverse function as a MOBIUS ribbon from the continuous deformation of an original circle in link to the source point?

The universe would be a unidirectional multiverse.

And on the vibratory plan, does space behave as a labyrinth? That would be the logic of perpetual loops, the system doing a loop onto itself and that would justify the original source point obeying the format of the recursive schema.

For this reason, we must approve three main principles:

- All is circular
- All is submitted to waves
- All is recursive

2. Necessary Preamble to the Constitution of the Existing

The primal condition for the creation of the universe is energy. Without energy, there is no living possible.

Every living form being maintained alive with energy, we shall question, thus:

1. Is the universe as any living body, destinated to reproduce or multiply itself? Is it in constant expansion and then due to a decrease? We could compare the evolution of each living form to a simple elliptic curve with optimal foci from which a decreasing phase would be initiated. The concept of evolution and involution. (Remark: According to the fractal principle, one element of the multiverse could be removed from it all?)

2. There would be an alpha and omega, a beginning, and an ending? We could presuppose that the universe auto regenerates itself by giving life to other celestial bodies while provoking the programmed extinction of some other bodies. From that disappearing, would another living body be initiated or due to be created. It is the concept of predation and of the natural selection of the living.

3. From the moment when the decline point would be culminating until getting a null value, it could adopt an inversed evolution curve in the negative values. This could confirm our notion of the sequencing of time into a recursive loop. The system would build

itself in an inversed proportional manner to the first evolution curve? Perfect symmetry of the curves?

3. By Referring Ourselves to Our Presuppositions

- Determinism: By reasoning with analogy, would the growth be modified in function of the parameters of any anterior information source: a cosmic corrective or a scale of added value? To the seeking of harmony and of an optimised growth combination.
- Randomness and probabilities of determinism: The coefficients of discordance and variation could act on the curve evolution, according to a series normally predictable? Anticipable?
- A series of aleatory events tending to...we could not suppose then the symmetry of the curve.
- Unless we believe in a genesis memory, of an inductive synergial type, based on the pre-existing structure. Synergy integrates external and internal inner phenomena into the system. Maybe we could reach that ultima objective of elaborating by ourselves. Of being the auto-creators of a system to the identical.

- Analysis of universal living, elliptic foci of decreasing

1. Through these examples, understand that the universe has a similar development to any living body, that it

has to be fed, it disintegrates and that it is in concurrence with the other expressions of the living. With a consequence: **Predation is original (present from the origins) and organisational.**

2. It is question of a monadic law (cyclic) to apply to the living scale. In order to do so, it would be necessary to establish a credible time reference. Not to have to use TEN at a power of 30, 50 or more. Feeling able to start from an original point that would be the correlational preamble and apply to humanity a time scale of 10-25(infinitesimal).

3. **To situate: It -1 before the universe creation.**

4. **To define a feeding law about the living and not to describe a cycle or alternance, what would be impossible to calculate is the too huge importance of the parameters of entry and exit into the fractal system.**

To finish:

- It would be inductive reasoning, as logical as to give evidence:

- Of a preamble to the existing. It is an action of concurring events with a resultant light. The notion of motion is not defined by an abstraction, but as the intended go through from a determined state to another whose determinations are unknown. In reason of its primary determination, any move would be an intent and hold some end limitations.

- A universal predation, an auto participant of evolution.

- A distributive system of waving or undulating energy.
- An alternance and not a cycle, considering the system as an energy provider and by adopting vector symbols.
- Some internal factors of the system as an infinitesimal move of the inertia centre, variation vector of the motion quantity, modifying the system codification suggested.
- Some external factors can modify the mass and gravity, inducing de facto a consequent sub-modification of our system codification. Except by creating an aleatory variable and attribute an effect ratio of 1% max in link to the considered masses. Fex $G(V)/M$. These two factors have the same inner value.

4. Proposition

$$A|x10^{75} = V(E/T) \times V (FxG/M) \times V(FxG/Ci) + Va (1\%)$$

D. The Universe

1. Principle

Uni versus, in the Latin language, signifies, so that to form a: ALL. To this signification, shall we attribute any religious meaning?

We are going, hereafter, to develop the concept of the GA, but we are starting from the following postulate:

The GA is the sum of parts, integrating numerous subparts. It shall be thought into terms of MULTIVERSE and of complex systems, as we are going to look at our subject through the concept of a systemic organisation.

2. Developments

Then, if the GA gathers numerous universes while being the GA, we should think that:

- If we admit the existence of a multiverse, describing it as intricating, the codification remains similar. The GA integrates the subparts of all in a codification where the system composed of many subsystems forms a unique system.

- Nevertheless, if one of the subsystems should be modified, would it affect it all? Does the system integrate any internal parameter that can remedy the deficit of its subsystems?
- But how can we modelise the All without thinking that each subpart is part of a whole and may have, in that sense, an equal value to the others, or a value in link to the others. As we do not know which value could be attributed to the Multiverse, we cannot attribute any relative value to the corresponding real ones within the whole.
- We shall accept the way of thinking according to which each subset is One into the All, forming a unique set within all, as our Earth, unique but integrated into a whole. We cannot add them but multiply them, as they have the property to be interconnected. They are indivisible, can reproduce or be reproduced and are multipliable.

Knowing that in function of the universe expansion, some universes can be subtracted from the All, but the All is a reproducing set. It multiplies more than it aggregates or adds (fusion).

It is then right to think:

By multiplying the set of the subparts, we obtain the unique multiplicity of the All. This is especially true if each subpart has the same value.

Of course, we can legitimately think of each part as a Plus into the All, but it is important to consider each of those parts as participating in the reproducing of the All. An evolutive All which fusion or of which an element can be withdrawn. But it maintains itself. This is the logic of the system.

3. Other Developments

We should assume that:

Within that global and globalising system, each one functions in an autonomous way with an entry and an exit. They are systems into great ones.

If each system or set (S) has its own functioning within the All:

$$S= S1+S2+S3+S4...$$
$$Or\ S/S2=S1$$

Each system has a common subpart.

4. Referential Glossary

We have to determine a glossary of reference.

1. Then, the multiverse would be interpreted in mathematics as an **Interverse.**
2. Then, the subpart to each set will be called the **Intraverse.**

5. Other Suppositions

S we do not know the set total value
To calculate 1, would be called: S, the All system.

Each set will be attributed an alphanumerical value that can be declined to decrease to the infinite by attributing new successive primal values if the value has to correspond to a universal scale.

Then, for example, A1' symbolised with an arrow vector to consider SetA1' from a conscient and subjective scale.

(S)= A1+B1+C1+D1+E1...

We can suppose a hyper translative system going from one subsystem to another.

Then, assume that each system is in link to the other, there would be a common subpart to each, which we have already named Intraverse.

So that:

S=A1+AB1+B1+BC1+C1+CD1+D1...

Every Intraverse is an internal intermediary opening, an inter the system common space, in a general Interverse Set. It would successively access the auto-organised core system.

If we position from a more mechanical view of the system, there would be a CORE SYSTEM, an INTERNAL ENERGY, an IN/OUT ACCESS, an INTRASYSTEM...

This system will itself be fragmented and fragmentary. It functions as a lens, as a curved surface that makes the set of rays that goes through either converge or diverge. By considering this alternative, it would be semi-reflecting, served by a continuous energy influx with an added value at the energy impulsion moment.

We could define it as transversal to the other systems and in traversable(able). The transportability of one system to another may include some regularity variables. It is submitted to the energy influx of any In/out accesses. At this stage of our exploration, we cannot envisage any variables. However: If E: Energy

Intraverse/intertime+intratime = F(E), interval where the energy flux is at its peak. We could think that the distribution of energy into that Intraverse, observes the deviation of a light ray going from an open space to a closed system in other parts.

The system, observing, by itself, a diffraction at the neighbouring of any opaque bodies and of the corpuscular and undulating energy as defined in the previous chapter.

6. Other Suppositions

The Intraverse is luminescent. It emits some cold light after receiving a ray. Is it quantum imbrication?

The quanta being: The manifestation of light into some little discontinuous quantities.

It would be a system differential of balancing, to pass from one system to the other. At this stage, we do not know if it induces some decompression stages. As every system is submitted to universal laws and to any external influences, modifying its auto organisation, into permanent adjustment, authorising the system maintaining.

We have named this common space into two universes the Intraverse. This SAS would authorise the indirect access to the System, and we have defined the Multiverse as the Interverse, for calculation purposes.

The differential is integrated into the system and functions according to the principle of the contrary

extremes: Evil twins to any elementary particles. They fusion into the same set. The original darkness was not an opposition to creation, but the necessary differential to the creation of its contrary. Without differentials, energy cannot be provided. This comes with the fusion of two opposite charges. It is the same with All.

E. Motion

1. Principle

We have chosen to envisage the motion concept in link to the two preceding chapters, as we should specify Motion as circular and continuous (as the celestial elements), as a move without any determination.

It is not a discontinuous move that could possibly be symbolised into a line.

Then, two notions of motion oppose themselves. The indeterminate move of the cosmic elements and the discontinuous move of any living body. If walking is going from one point to another and we could, at this title, considerate the determination of a move, we would objective its possible ending and discontinuity. Determination is going from one point to another. Move does not refer to any global action, but to a system mechanic.

- Hypothesises

- Could we think of symbolising motion into a number that would code an always redefining infinite? We need a new maths sign integrating the notion of universal systemic proliferation, as it is like a living body destinated to reproduce itself.

-
- As: it could be possible that this expanding universe would grow till reaching the right balancing to the whole, beyond which the system will be auto limitative or a balancing base that would help maintain it as such.
- As: if we adopt the idea of a Source point with some inertial and gravitational properties, we assume that Source Point once defined, to be constitutive of an alignment and of a determined revolution move in link to an axis situated into the Source Point (the objective of our prototype machine would be to define any potential interactions from that point).

Then, the circular move adopts a consensual direction to the living. The presupposed direction is established as unidirectional.

<u>By referring to the recursive time loop, we have established an internal loop intrasystem. In the notion of space, if the system functions in an auto-loop, it has some mechanical properties.</u>

It would have a double sense of internal rotation, that is, the principle of a recursive system.

We can observe that it is a principle that maintains the ISS in a balancing state.

2. Other Suppositions

If an axis rotates in one way and another in an inverse way, there is a force that is exerted onto the system to give it some stability.

However, the universe being constituted of some intricated systems one into the others, they auto-influence themselves according to some external and internal factors.

Then, if the system is conceived to remain stable, the totality of the systems and subsystems is presumed to be unstable.

The idea of motion that we are developing has more to do with A GENERAL DIRECTION than with the idea of continuous and discontinuous lines. To us, the system turns according to a ternary principle.

We have kept in mind, from the beginning of this discussion, the concept of odd parity by creating a figure assembling, without which it will be impossible to code. For example, (-1, nought, +1). Refer to the chapter.

Nothing seems to be thinkable out of the ternary system. From the concept of time, until that of motion, that cannot be summarised to a continuous or discontinuous line, as it obeys to the universal circularity.

F. Calculation of an Original Source Point According to the Systemic Principle

1. Principles

The Source Point will be what has generated the system and helps to maintain the system. It auto-regulates each of the system elements and any modification into the balancing of any of them, representing a threat to the balancing of the whole, even if that modification is infinitesimal and inherent to the only misfunctioning system. We should not forget that the chaotic element is thought of within a universe holding a principle of predation; it would then be repelled from the system as it was integrated into and as we have tried to conceptualise.

2. Developments

The source point authorises to get back towards the Instant, when the system constituted itself and we will be trying to understand and establish:

The moment when the misfunctioning system will be repelled of the All, either by intra or inter destruction.

The way our system has been generated, so that we join another source point in a spacetime functioning in parallel to ours, and that we have evoked in the chapter about time.

From the moment when the Source Point generates a system, it develops an alternative in parallel as a consequence of the conceptualisation of the system. Remember the notion of differentials.

Then, the constitution of S1 consequently gives birth to S1', an asymmetric universe with two same sides in link to the source point. An internal double to the system. The source point underlines, by the light waves at stake, the recursive loop, that it generates the possibility to get back to T-1, as regards to T+1. In other terms, the chaotic element hides a principle of quantum doubling (under the form of an area called reflexivity area), where time observes the same retrograde demarche face to some similar alternatives but where the projected actions could have interfered differently on the elementary system S1'.

3. Principle of Perennation of the System and Its Rebalance

The Source Point projecting a beam of light (in accordance with the definition of a bundle of light rays) in a triangular shortening way from its base to its summit, it would hold in its other extremity, two sides of equal length, composing a trapezoid whose superior angle would be more than 15°, authorising a gravitational waving reflexion by the intervention of the central canal of gravity of the Source Point. It is a question of the reflexivity area, hollow zone, carrier of impulsions. It is a central zone, in straight link to the system maintenance.

This area would permit the dispersion of light waves; it is that set of prisms essential to the continuity of the system. It is invisible because of its positioning in the beam of lights.

The construction of a prototype would authorise some calculations about the interactions between the reflexivity zone and the triangular base of the beam.

Is it an autonomous zone, functioning in sub-hand to the system, possessing some decompression stages to access it. As a sub-hand, it would keep the same characteristics to the light and in the change of direction of the waves, while we already presuppose their refraction, in link to the proper nature of the zone.

Is it constituted of hollow matter, of constantly distorting matter, fluid and of a solution of unstable gases.

As it is a reflexivity area, it has the role of transfer/equilibration.

It functions as a compression zone, reflecting energy back, but it is not in itself an energy zone.

It observes a quasi-symmetry to the source point.

4. Effects

- Effect: gravitational on matter. The supersymmetric particles carry impulsions and energy.
- Global effect on the multiverse.

- Parallel world of dark matter. From a hollow zone to a virgin zone. Mirroring effect. 2 structures with a mirroring effect. 1

That zone of reflexion is a hollow zone of transfer, inexistant from the point of its mass, where density is so weak, that it is a zone with prism action. A zone of transfer. It could be constituted of less dense dark matter.

The source of light is conical, and its summit rejects the infinite. It was like the light of a torching lamp that would be projected into darkness.

It requires some water for beam conductivity, some air for combustion and some friction to the liberation of a gas spark. This system is an autonomous indispensable system whose alignment is not centralised.

It is at the initiative of all universes, but its alignment depends on the stability of dark matter. In other words, if that dark matter observes a behaviour, such as absorbing some other dark holes, the base displaces itself. It occupies a central position but can be closer to one universe than to another.

As the location of the base to the source point towards another universe implies that the beam and the gravitational waves that are emitted have not the same intensity, the living modifies itself.

Could we then presuppose an anteriority to the existing multiverse?

- A pre-existing to the existing is necessary and the GA has not been created from nothing, but it is an evolution of matter. Matter is evolutive. The maintenance of its equilibration would rely on diffraction into the inter systems. The circulation of energy is not translative within our system.

G. Twin Aspect of the Universal System

1. Principle

We then established that the universal system is an ALL and assumed the source point to be potentially reproducible. It would function as a beam of light reflected on a mirror side. From an original source point, successive creation of other source points.

What we see of the universe is like a projected image of universes quasi-similar, as they cannot be similar if we consider the internal predation within the different systems and the adaptative evolution of the system.

We will be calling this predation principle Predaction.

It authorises the different systems to evolute according to their positions into evolution itself and their position within an arborescence. It authorises a conscient and organised development according to a plan established by each subsystem.

2. Developments

What could seem the most established would be the twin aspect of our universal system.

If we start from the principle of the living and observe the living, then its duality appears. It has a dual nature, of the same type; a differential system of equilibration, dedicated to maintaining all.

The living, which we are components of, integrates in itself the universal system. It is a differential system of equilibration.

To create a continuity to the living, two energy influxes are to be accreted so that to create an extra system =+

The living is a system All into the universal system.

Universal: it seems to assert the unicity of the system. We have admitted the set to be One and All to be in that One. Everything is unique and multipliable. The set of all those Ones is constitutive of the living and of the universal All.

Their functioning as to energy is different. The multiverse functions on the principle of quantum intrication, supposing the manifestation of light through discontinuous quantities.

If the universe has always existed and evolved to the Interverse that we know, it has observed an energy continuity. The primal and basic requirement of the universe and the living being energy.

A continuous link only. The system evolutes: it has generated a BIS repetita system that has evolved differently according to gravitation and energy.

Then, our galaxy evolving into the Interverse in an autonomous way, while being linked to the whole, cannot maintain as it is. It is evolutive and predactrice to its turn.

143

Earth, which shelters the universal living that we know of, must integrate the Interverse and think itself into an ongoing process that is processed, and which is processing, by the whole.

The maintenance of living depends on our capacity to create or recreate a source of autonomous energy if the system thinks itself as unique.

If it thinks as One into the GA, it is connected to the original source points and can be re-programmed according to the original conditions of gravity, stability and energy. Quasi-similarity of the sub-originated system.

A reinitialisation of the original system. Or how dark matter generates and emits energy, how does it contribute to the Interverse stability.

Indeed, we have to interest ourselves in this Twin aspect. It would be a sub internal system and, by thinking of circularity, could apply more generally to matter, matter coupled to energy organising the living and any favourable conditions to its renewing or continuity.

How can we not think of the fact that an elementary particle could have intersystem mobility. We are not able to predict if it obeys a certain determinism, or if its flux can be anticipable. Once split into two particles, how will they evolve?

They will not be submitted to the same external factors but will develop themselves in function of their principle of adaptation, which is the recurring principle of the ALL.

Do those differentiated particles keep in themselves some original memory, we mean: have they got the possibility in spite of different factors of mobility and evolution to fusion once again, to repel or auto-destroy each other?

It would induce that some similar particles could have evolved differently and have adapted an intra-system that does not answer to the same schemas of development but which, having kept some of their original memory, could attract once again. This would lead to the idea of a particle regrouping two adaptative schemas: to repel what is not recognised? To predate? Fusion?

Which way would predation be exerted?

There is no evidence than a more elaborated particle has more capacity to absorb another, because of its elementary constitution and of its zero mass.

What links the two particles to their creative memory? And to us, what has been created is de facto divisible into one subpart and into another? However, there is the original link.

This link is continuous in its motion and discontinuous in its adaptation seeking.

However, does this link authorise any predictive perspectives of evolution?

In other terms, we can envisage:

A particle $X/2 = X^{2}$'

And a set of split elementary particles.

There would be a differentiable state:

$S(X/2)$ tending towards $S(X^{2})$

Two differential sets? $S'X')$, $S(x')$, $S'(X)$, $S(X')$

and that each set has developed an autonomous functioning, not in adequation to the original set:

Can we deduce the original set of the split particle:

Eo $(X/2)$ obeys cyclic law?

Can we think that this original patrimony would act on these particles and activate an original link?

Does this potential conservation of the original properties coupled with the evolution of the particles into more or less favourable conditions and medium of development authorise us to anticipate some laws of evolution of those particles?

For example, establish a cycle, a momentum?

Where do the particles get back to the original point?

Mo + evolution

Which objective for these calculations?

A system that will be a set of elementary particles.

That following universe would be issued from the original universe? Could two split universes evolve differently or respond to a cycle to join their original point at the end of a term of evolution?

It seems much more complex.

If: p- as the incidence of move within ternarity

M1x p x p-1 p- +lim2 = Po x T x E/C

P in blue as a vector representing the symbol of move

3. This Subject Interests Ourselves

It is a question of being able one day to anticipate any attack on our system, any evolution that could threaten our All. To be the masters of our own fate.

But in a world that is constantly in motion, which is the right solution?

Being able to join some other temporalities? Find the calculation for the derivations of a bridge point into spacetime?

Know more about the cosmic living cycles? But how could we without any previous references?

I have already asked you: we have to look at the invisible and change our way of envisaging the universe. Then, a solution will appear. It is the truth that it must appear.

As a consequence of a conscient and organised evolution that wants to perdure.

Ostensibly, one major stake emerges from our suppositions:

Developing a better understanding of the multiverse and seeing Earth as a part into a whole, think it as a sub-part of a system, and consequently of the GA.

Given this, the Earth is acted and interacted, and the system has an action effect within the system. Some internal and external factors, some aleatory ones, interfere in permanence.

Nothing being ever constant or indefinite, the system adopting some ternary laws of circularity and recursivity, the present time, should but only be a concept, not a reality.

We, ourselves, are due to live within a certain time period.

If we adopt the idea of the universal well founded, we could think that a plan governs the universe according to a conscience. The only capable of eradicating earth should be the starting point, of the system itself on reasons of general interest but considering the living as the original cause of creation, this hypothesis is naturally excluded.

If we do not believe in determinism but in a succession of aleatory events, how will it be possible to find any justification for such an ideal defined objective?

To us, there has been a first intention, that wants the best. We have called it SUPERIOR INTELLIGENCE.

It has named us, in conscience, to, in our turn, bear an Intention...

Humanise the indivisible invisible and carry the VERB, high, as a unity of action.

Fusion the universe and the living into the same entity.

It requires some necessary wisdom to come to that better understanding of the living that does not interfere with the questions about determinism.

Comes the time of a new alliance between humanity and the intention.

In that specific period, shall we remain confident in our destiny or do nothing?

Knowledge elaborates itself against any anterior knowledge. (Bachelard)

Knowledge as a liberating concept to humanity.

Through this novel, we wish to create a debate with a subobjective that is the necessary preservation of the existing balance between the multiverse and the living.

Obtaining the accurate type of information, mastering and developing future alternatives.

We also evoke some different options on the functioning of the celestial system and the living, some hypothesises including a multitude of parameters, and some calculations with various variables.

But we just tend towards a unique objective:

Our first and unique commandment is and will be the LAW of LOVE; it is the only law that must command to humanity. (Faith Act 1.1)

Printed in the USA
CPSIA information can be obtained
at www.ICGtesting.com
LVHW021351051023
760085LV00064B/2029